Retreat from a Rising Sea

Retreat from a Rising Sea

Hard Decisions in an Age of Climate Change

Orrin H. Pilkey
Linda Pilkey-Jarvis
Keith C. Pilkey

Columbia University Press
New York

Columbia University Press wishes to express its appreciation for assistance given by the Santa Aguila Foundation toward the cost of publishing this book.

Columbia University Press
Publishers Since 1893
New York   Chichester, West Sussex
cup.columbia.edu

Library of Congress Cataloging-in-Publication Data
Names: Pilkey, Orrin H., 1934– author. | Pilkey-Jarvis, Linda, author. |
Pilkey, Keith C., 1965– author.
Title: Retreat from a rising sea : hard choices in an age of climate change /
Orrin H. Pilkey, Linda Pilkey-Jarvis, and Keith C. Pilkey.
Description: New York : Columbia University Press, [2016] |
Includes bibliographical references and index.
Identifiers: LCCN 2015034502| ISBN 9780231168441 (cloth : alk. paper) |
ISBN 9780231541800 (e-book)
Subjects: LCSH: Coast changes. | Sea level. | Shore protection. |
Global warming. | Coastal zone management.
Classification: LCC TC330 .P55 2016 | DDC 333.917—dc23
LC record available at http://lccn.loc.gov/2015034502

∞
Columbia University Press books are printed on permanent
and durable acid-free paper.

This book is printed on paper with recycled content.
Printed in the United States of America

c 10 9 8 7 6 5 4 3 2 1

COVER IMAGES: Justin Leibow (*leibow.com*), Xeromatic (*lifeofpix.com*),
Oscar Keys (*unsplash.com*)

COVER DESIGN: Randall Bruder, Good Done Daily

Dedicated to
Sharlene G. Pilkey

What'cha gonna do, what'cha gonna do
What'cha gonna do, what'cha gonna do
What'cha gonna do, what'cha gonna do
When the land goes under the water?
Can't go east, Can't go west
What'cha gonna do, what'cha gonna do
Black tide's coming it'll do the rest
When the land goes under the water?
Can't go north, can't go south
What'cha gonna do, what'cha gonna do
Can't go swimming to a big whale's mouth
When the land goes under the water?
The land is broke, the skies are too
What'cha gonna do, what'cha gonna do
Too little warning when it comes for you
When the land goes under the water?
What'cha gonna do, what'cha gonna do
What'cha gonna do, what'cha gonna do
What'cha gonna do, what'cha gonna do
When the land goes under the water?
What'cha gonna do, what'cha gonna do
What'cha gonna do, what'cha gonna do
What'cha gonna do, what'cha gonna do
When the land goes under the water?

BÉLA FLECK AND ABIGAIL WASHBURN, "WHAT'CHA GONNA DO"

Contents

Foreword

"For the past two centuries, two trends have been steady and clear around the United States. Sea level has been rising, and more people have been moving closer to the coast," stated NASA scientists. The steadiness and ineluctability of these two trends have been observed and documented worldwide as well by scientists and international organizations. Such migration of people to coastal regions is common in both developed and developing nations, and according to an IPCC report, utilization of the coast increased dramatically during the twentieth century, a trend that seems certain to continue through the twenty-first century

Corroborating statistics from the United Nations indicate that half the world's population lives within 40 miles of the sea, and three-quarters of all large cities are located on the coast. Furthermore, most of the world's megacities, with more than 2.5 million inhabitants, are in the coastal zone.

The attractiveness of the coast, leading to the increasing rate of seaside settlements, can be explained by the economic benefits that accrue from access to ocean navigation, coastal fisheries, and tourism and recreation, and has resulted in the disproportionately rapid expansion. Indeed, unbridled urbanization on risk-prone areas along beloved, yet hazardous, shorelines is placing populations in danger zones. Under mounting evidential reports, scientists are warning us that, due to climate change and rising seas, storms are expected to be stronger and fiercer than ever before. Reminders of seaside-living risks abound too tragically in the news, worldwide.

The acceleration of human migration toward the shores is a contemporary phenomenon, but the knowledge and understanding of the potential risks pertaining to coastal living are not. Indeed, even at a time when human-induced greenhouse-gas emissions were not exponentially altering the climate, warming the oceans, and leading to rising seas, our ancestors knew how to better listen to and respect the many movements and warnings of the seas, thus settling farther inland. For instance, along Japan's coast, hundreds of so-called tsunami stones, some more than six centuries old, were put in place to warn people not to build homes below a certain point. Over the world, moon and tides, winds, rains and hurricanes were symbiotically and naturally guiding humans' settlement choices.

Modern humankind appears to be the only species on earth whose propensity is to migrate its habitat counter-intuitively, ruled solely by will, preference, greed, and, most dangerously, a sense of technological and engineering invulnerability against nature's changes.

With our children's future in mind, we must reconnect with our innate memory of the risks. We must be humble and accept scientifically corroborated facts.

We must be reasonable and become malleable to nature's evolution.

We must be wise and build farther inland at higher elevations. We must be courageous and accept the need to retreat from the shores.

Insanity has been defined as doing the same thing over and over and expecting different results. We must cease the insanity, as the seas are rising . . . ineluctably.

The Santa Aguila Foundation is proud to have helped make *Retreat from a Rising Sea* possible. The Foundation is a U.S. nonprofit organization dedicated to the preservation of coastlines around the world. It was created after the Foundation's founders witnessed the destruction of the beautiful beaches of Morocco by sand mining. Since then, the Foundation has focused its energy on global coastal issues and education. Our education efforts include the management of the beach website coastalcare.org and the publication of four books authored by Orrin Pilkey: *The World's Beaches: A Global Guide to the Science of the Shoreline*, *Global Climate Change: A Primer*, *The Last Beach*, and *Retreat from a Rising Sea: Hard Decisions in an Age of Climate Change*. We also brought our support to the award-winning documentary film *Sand Wars*.

We hope that you will enjoy this book and take as much pride as we do in defending and protecting our coastal environment.

For further information, please visit www.coastalcare.org.

THE SANTA AGUILA FOUNDATION

Preface

In 1969, Hurricane Camille roared ashore on the Mississippi coast with wind gusts in excess of 200 miles per hour, producing an astounding storm surge of more than 20 feet. Possibly the strongest, though not the largest, hurricane to hit North America in the twentieth century, Camille was a Category 5 storm.

My parents had retired to Waveland, Mississippi, in the early 1960s. Their house was situated at an assumed elevation of 20 feet, presumably high enough to be safely above any expected storm surge. At one point, my engineer father did a hand-level survey from the shore to the house and found the house to be at a baffling 13-feet elevation. He was puzzled because before buying, he had examined city maps to make sure that their house was at 20 feet. He decided that he must have miscounted some of his level sightings and didn't pursue it further.

At the height of the storm, my parents' house flooded, filling with 5 feet of water. My brother and I spent several weeks there, cleaning the house and recovering their life. The mystery of the elevation was solved a few weeks after the storm when my father reexamined a city map and discovered, written in fine print in the legend, the notation that sea level was 6 feet, meaning that the zero-elevation line was 6 feet below current sea level—based on the earth's geoid. The *geoid* is the shape of the ocean's surface if only the earth's gravity and rotation are in control and there are no winds or ocean currents.

There is no discernible reason to base city-planning land elevation maps on the geoid. No doubt, few citizens of Waveland knew that sea level was actually at 6 feet elevation. Probably the city officials had misinformed them in order to encourage more development, expand the tax base, and increase prosperity. If so, the cost was lost lives and destroyed property. This was my first lesson in the length to which coastal communities would go to in order to encourage development in dangerous areas.

Hurricane Katrina arrived with true terror 36 years after Camille, in 2005. By this time, my parents were dead, and another family had long occupied the house. Katrina's storm surge was a remarkable 28 feet. The house, along with the entire seaward part of the town, was wiped clean. Along the five city blocks between the beach and the concrete slab that once held my parents' house, essentially nothing remained but other concrete slabs and tough live-oak trees.

A few months later, on a visit to Waveland, I saw signs on various concrete beachfront slabs offering home sites for sale. Unstated in the real-estate sign was that some of the slabs, which were for sale for $700,000 to $800,000, had once been occupied by houses destroyed in both Hurricanes Camille and Katrina.

The lesson here for me was that the drive to live next to the beach and the push for profits will be an almost insurmountable hurdle to overcome as we respond to the inevitable rise in sea level.

It was this family connection that inspired the production of this book about the necessity of retreating from the shoreline. The level of the sea is rising, the rush to develop near the beach continues, the size and number of beachfront buildings go up, and storms are intensifying—all of which means that the magnitude of future disasters will increase. Had the city of Waveland decided to implement a retreat policy in 1969 after Camille and prohibited the reconstruction of destroyed houses, the damage from Katrina would have been much, much less. Instead, in much of the South, there is a strong undercurrent of skepticism about global climate change, including the reality of rising sea level. Why bother with retreat when the future of the sea level is uncertain and especially when there's money to be made?

Another motivation for us to write this book was the gradual diminishment of state-level coastal management programs in the twenty-first century as good regulations from the 1970s and 1980s (such as the prohibition of building beach-destroying seawalls) were compromised after influential people's houses were threatened by erosion. In the case of North Carolina, the state's once courageous and pioneering coastal management program has virtually collapsed on itself, and the prospect of sea-level rise has been ignored and denied.

The Plot

The initial chapter in our book lightly touches on our main conclusions. This is followed by a study of the mechanics of sea-level rise, how it is measured, and why it varies from place to place along the world's shorelines. To understand the credibility of sea-level rise projections, we explain the workings of the Intergovernmental Panel on Climate Change (IPCC) and then look at the future of three cities and one country, a study of great contrasts. Miami and New Orleans are the two cities most vulnerable and least prepared for sea-level rise in the United States. Both are doomed in the long run. We compare the United States' best-prepared city, New York, with the world's best-prepared country, the Netherlands.

The plot moves on with discussions of the role of geology, politics, societal wealth, and culture in determining the likely response to sea-level change. Atolls and small Alaska Native villages are in immediate danger, and retreat is already under way. In less-developed countries where houses are sometimes built to be mobile, retreat has been under way for many decades. In wealthy countries, sea-level rise is often ignored, as it is in South Florida. But in the Netherlands, it is not only recognized, but a carefully planned response is already being carried out. The British plan to abandon some lowlands, and a number of states and countries have prohibited or restricted seawalls. Other topics that we explore include

- The infrastructure that is threatened by sea-level change, including landfills and toxic-waste sites, water-treatment systems, landmark buildings, and energy facilities.
- The legal framework for subsidizing community responses and the failure of staying in place (mitigation) rather than retreating.
- The major humanitarian crisis that we face around the world as environmental refugees flee to higher ground. Thousands of residents of Arctic shoreline villages, a million Pacific and Indian Ocean atoll dwellers, and millions of delta inhabitants will soon be environmental refugees.
- The huge problem of land loss and environmental refugees fleeing to the cities or other countries, conceivably a cause of social strife and even wars.
- The climate change deniers and their corporate supporters, with an emphasis on those that dispute sea-level rise.

In the final chapter, we briefly recount the shoreline history of the United Kingdom's Holderness Coast to make the point that we are dealing with a long-term problem. Since the invasion of the Romans about 2,000 years ago, the shoreline there has retreated by 2 to 3 miles, and 28 villages have fallen into the sea and now reside on the continental shelf.

ORRIN H. PILKEY

Acknowledgments

First and foremost, we acknowledge with much gratitude the support of Eva and Olaf Guerrand Hermes and their Santa Aguila Foundation, discussed in the foreword. They are friends of the beaches (and us as well) and staunch supporters of the effort to preserve beaches for future generations. Among other things, they are resolute opponents of the mining of beach sand, an important but often unrecognized factor in beach destruction worldwide. They also have started the foremost coastal/beach website: coastalcare.org. The website manager, Claire Le Guern Lytle, furnished us with much information (and insight) for the book.

We have benefitted from discussions about responding to sea-level rise with many colleagues over the years. Hal Wanless, Paul Stout, and Al Hine assisted us in understanding the future of Miami; Joe Kelley and Rob Young helped us through the complexities of New Orleans.

We are indebted to a large number of people who have furnished us with information, ideas, and suggestions. They include Andrew Cooper, Bill Neal, Andy Short, Alex Glass, Peter Haff, Fred Dodson, Frank Tursi, Rob Young, and Miles Hayes. Bill Neal did us the privilege of reading the entire manuscript at an early stage and making numerous suggestions and comments. Andy Coburn provided much information about the world's beach-replenishment programs. Tonya Clayton sent a raft of media articles that boosted our understanding of the public's view. Thanks especially to Norma Longo, an invaluable research and editorial assistant and able copy editor, who managed to organize the three authors to prevent much overlap and duplication. The Nicholas School of the Environment supported us throughout the writing process. We also benefitted from financial support from North Carolina State Representative Pricey Harrison, one of the state's leading proponents of a sound long-term coastal management policy.

We really appreciate the patience and support of our significant others: Jim, Melissa, and Sharlene. Unfortunately, Sharlene passed away as we were finishing the writing, and so we dedicate this book to her.

Retreat from a Rising Sea

1

Control + Alt + Retreat

The coast is crashing with the rising sea. There is no question that the sea will continue to rise and that the rate of rise will accelerate. Even now, in our lifetime, sea-level rise is affecting humanity as both Inupiats along Arctic Ocean shores and dwellers on coral atoll islands in the Pacific and Indian Oceans frantically seek new home sites. Delta dwellers on the Ganges, Mississippi, Niger, Yangtze, and Nile Rivers also are at great risk because the sinking of the deltas enhances sea-level rise and flooding. In Miami Beach, Norfolk, Annapolis, and Queens, some drivers worry about saltwater rusting their car axles as they plow through saline waters on low-lying streets when high tides arrive. The flooding problem is well known in a number of other cities around the world, including, most famously, Venice, Italy. The world's immediate future holds a horde of environmental refugees who must escape as the sea rises: thousands of Inupiats, a million atoll residents, and hundreds of millions of delta inhabitants.

Why Has the Sea Level Risen?

Our dependence on fossil fuels has in large part brought us to this place, causing a chain of events that warms the atmosphere, which in turn warms and expands the oceans, melts glaciers and ice sheets, and consequently raises the seas. The sea level has changed throughout the earth's long history, rising and falling and leaving behind a record for us to read. *Global variations* in the sea level are due to changes in the volume of water in the earth's oceans (primarily from expanding warmer water or melting glaciers and ice sheets) and very gradual changes in the capacity of the ocean basins (*global tectonic changes*). *Local and regional variations* in the level of the sea take place with the sinking or rising of the land due to the following:

- Earthquakes and other crustal movements
- Formation or melting of ice
- Extraction of water or oil
- Sediment compaction on deltas

In addition to changes in the elevation of the land, variations in the direction and intensity of ocean currents can cause sea-level changes, which is why the sea level is 8 inches higher on the Atlantic side than on the Pacific side of the Panama Canal.

In less-developed communities around the world, sea-level rise is a multiplier for underlying stressors and vulnerabilities, including poverty, pollution, isolation, and overpopulation. Inundation during high tides and storms degrades agriculture along the margins of bays, killing coconut palms on tropical shorelines everywhere, ruining drinking water supplies through salinization, and forcing communities to move.

The water's edge on all types of coastlines from the Arctic to the equator is advancing at rates that will surely pick up as the ice sheets in Greenland and the

The water's edge on all types of coastlines from the Arctic to the equator is advancing at rates that will surely pick up as the ice sheets in Greenland and the Antarctic break up. Greater shoreline erosion in the warm climes of the more-developed countries will lead to continuing large expenditures in futile attempts to hold shorelines in place and to preserve the all-important beaches. But saving beaches is a battle that won't be won.

Antarctic break up. Greater shoreline erosion in the warm climes of the more-developed countries will lead to continuing large expenditures in futile attempts to hold shorelines in place and to preserve the all-important beaches. But saving beaches is a battle that won't be won. Ultimately, beachfront property owners and communities will favor the preservation of buildings over beaches. In fact and unfortunately, within a few decades, the battle to save the beaches in developed beachfront communities will be lost.

There are so many inequities in this crisis. The developing coastal countries are beginning to demand monetary assistance from the larger polluting countries for the retreat process and compensation for their suffering. Island nations such as the Maldives and less-developed coastal countries like Bangladesh are demanding a halt or at least a reduction in the production rate of greenhouse gases. These nations believe (perhaps correctly) that those that contribute the least to the human causes of sea-level rise will be hurt the most. Unfortunately, even if we reduced the production of CO_2 tomorrow, it would have little effect on the rate of sea-level rise in this century.

The deniers of climate change and sea-level rise continue to have a voice that seems to grow weaker with each superstorm. But a closer look shows that the deniers provide a facade of credibility for a host of politicians who contrive to ignore the rising sea. Deniers have vested interests (mostly related to the fossil-fuel industries) in confusing us and thereby delaying regulatory action. After all, the cost of responding to sea-level rise will be immense, and as long as we delay a response, there's money to be made by fossil-fuel companies and coastal developers. But there's money to be made in the retreat process as well.

The deniers of climate change and sea-level rise continue to have a voice that seems to grow weaker with each superstorm. But a closer look shows that the deniers provide a facade of credibility for a host of politicians who contrive to ignore the rising sea.

Given the widespread, measurable evidence from tide gauges and satellite measurements that document the story, fewer skeptics now claim that the sea level is not rising. But a camp of deniers persists, arguing that the rise is both minor and temporary and that soon the sea level will drop again. Regrettably, politicians still commonly respond to questions about sea-level rise and its causes by claiming ignorance of the topic, saying they aren't scientists. Even though these politicians may not believe the findings of climate scientists, they are likely to believe the science of their doctors, dietitians, and meteorologists. Short-term deviations from long-term trends, such as the fact that in 2013 the extent of summer sea ice in the Arctic was greater than in 2012, are disingenuously touted as evidence indicating that the earth is cooling.

What is to be done with the many hundreds of miles of high-rises jammed up against the sea around the world, most spectacularly in Florida? To move all these buildings is not economically feasible, but even if they could be moved, a suitable place for them would be difficult to find. One could build seawalls that would have to enclose the islands completely and grow bigger and higher with time, but these would destroy the very beaches that drew the construction in the first place. Preserving beaches for future generations is a compelling reason to retreat in response to sea-level rise. As former Florida governor Bob Graham asserted, "This generation doesn't have the right to destroy the next generation's beaches." So for the sake of our grandchildren and great-grandchildren, we don't have the right to build beach-destroying seawalls to save our beachfront buildings from sea-level rise.

We can't begin to replenish (i.e., pump or truck in new sand) all of Florida's beaches. Furthermore, much of Florida's beachfront development is on permeable rock through which water will rise and flood behind seawalls or levees. Replacing cars with small boats might work until the waves from big storms roll through the community. Demolishing the high-rises would cost a fortune and produce a vast amount of water pollution, although modern demolition techniques could salvage and recycle large portions of the building materials. Thus, by the latter half of this century, much of the beachfront high-rise problem will be in our laps to solve.

We humans find it hard to grasp the magnitude of changes that are under way, especially when the deniers try to confuse us. Sea-level rise is at the forefront of the expected changes, and if the higher estimates of sea-level rise rates are valid, a true global human catastrophe by the end of this century is in the offing. Our perception of what is permanent or lasting will be challenged, even though nothing is happening with regard to the sea level that hasn't happened before in the geologic past. The recent disasters intensified by rising seas are not random events without underlying causes. Indeed, climate changes, including the frequency of extreme events, have advanced to the point that we can no longer predict future events based partly on what has happened in the past.

Even events on the scale of Hurricane Sandy rarely result in immediate significant changes in coastal development patterns. In the 1980s, many coastal planners and scientists were saying that if just one more hurricane hit, surely things would change, and we would start moving back from the beachfront and prohibit further construction in these extremely dangerous areas. Then along came Hurricane Hugo in 1989, and the idea of responding sensibly to the storm to prevent damage from the next inevitable storm was tabled.

At the time, South Carolina had a law that any beach house destroyed in a storm could not be replaced. This law was perhaps the most merciful and politically least controversial way to begin a retreat from the beach. But influential citizens with damaged houses howled, and the rules were changed so that if you could find the roof of your house, you could rebuild! When the dust cleared, Hurricane Hugo had become an urban-renewal project with many mom-and-pop cottages replaced by "McMansions" and even some high-rise condos. Several other post-Hugo hurricanes in the southeastern United States also proved to be urban-renewal projects when they should have been opportunities to retreat. These urban-renewal projects also have effectively priced out lower-income residents from the coast.

Why have we routinely brushed off any long-term response to natural disasters at the shoreline, all of which will be exacerbated by the sea-level rise? Certainly this is due in part to sympathy for those who have suffered loss, especially right after the storm. Too, there is an element of pride and nationalism that promotes rebuilding rather than giving in to nature. "We'll be back because we are Americans and we don't give up" is the cry often heard after a big storm has blown by. As one U.S. Army Corps of Engineers colonel explained, "We are not going to throw up our hands and slink away." The philosophy is that we will come back stronger than ever (but not necessarily wiser).

Greed and selfishness are often part of decisions to protect property at the price of beach destruction. In Southampton, New York, several beachfront billionaires are building massive walls to protect their individual homes, despite the community's opposition. In North Carolina, citizens from the "wealthiest island," Figure Eight Island, have succeeded in altering the state's regulation barring hard shoreline stabilization (seawalls). In South Carolina, Debidue Island, Figure Eight Island's equivalent, has obtained a permit to rebuild a seawall, which originally was against the regulations, and of course this will open the floodgates (pun intended) for other communities to circumvent wise planning. In Florida, new walls are rising with the approval of a state agency that no longer turns down permits for walls. Up and down the coast, we see what were good regulations in the 1980s being picked apart and slowly diluted in the twenty-first century. With increasing wealth feeding the rush to the beach, conceding the shoreline to nature has become increasingly unpopular.

Up and down the coast, we see what were good regulations in the 1980s being picked apart and slowly diluted in the twenty-first century. With increasing wealth feeding the rush to the beach, conceding the shoreline to nature has become increasingly unpopular.

What Can We Do About the Rising Sea?

In the short term, retreat is certainly not the only option at the shoreline, but it will be an integral part of our approach to rising seas.

Even in areas where we must adapt, some degree of retreat will be necessary. That is, we will have to abandon some homes, businesses, or lands or change land use in order to accommodate the floodwaters.

Many practical construction techniques already are in existence that can make coastal buildings more resilient. For instance, it is not uncommon for houses to be built on pilings, allowing floodwaters to flow beneath the living quarters. This idea of moving valuable living/working space above the potential flooding level can benefit cities as well. By using the ground floor and basements of buildings for nonresidential purposes, we can minimize storm damage. Instead, lower floors and basements can be used for parking structures that can withstand flooding with relatively minor damage, and the more valuable shopping or residential units can be safely situated above the floodwaters. For example, the Dutch have built an underground parking garage in Rotterdam that doubles as a water storage area to hold water during flood events. In addition, more and more parks in the Netherlands are being designated as storage areas for floodwaters that will be released once the flood conditions have passed. This type of accommodation on a wide scale requires some degree of retreat, though, in that buildings and houses must be torn down or relocated to make room for flood-control measures, most of which will become useless in a major sea-level rise. Wetlands restoration is another method that can and will be employed to help as a buffer against floods and storms, but this too may call for retreat in the form of buying out homes and businesses.

We in the United States undoubtedly will follow the Dutch example and construct large-scale dikes and levees to protect our cities. Massive dikes, too, can offer some cities the opportunity to expand housing and commercial enterprises, thereby reducing some of the cost of building large-scale protective structures. This appears to be on the verge of happening in New York, as discussed in chapter 4. Although the Dutch have led the way in engineering technology, which we also describe in chapter 4, they are turning away from their previous hard stance of not losing any land to flooding toward a softer approach of "living with water," that is, accommodating floodwaters in some areas while protecting others.

We can help curb inland coastal flooding during storm events as well through green infrastructure, such as living or green roofs that can help absorb storm water. We can improve water management to decrease runoff, though as we point out in chapter 2, sea-level rise presents challenges to managing inland floodwaters.

There are many creative ideas about how to accommodate the anticipated increased flooding associated with sea-level rise. The Dutch architect Koen Olthuis appears to be leading the way in the use of floating structures. Olthuis calls for letting "water in and even making friends with it." His amphibious architecture allows buildings to detach from the ground during flood events. Olthuis's company and the Dutch Docklands firm have built Greenstar, a floating hotel and conference center in the Maldives, and are constructing a similar floating hotel in Norway called the Krystall. H&P Architects, a Vietnamese firm, have designed a prototype of a low-cost floating house made from bamboo that is secured to pilings and sits atop recycled oil drums to allow the house to float during flood events. Similarly, Ann English's Buoyant Foundation hopes to retrofit Louisiana homes in flood-prone areas atop Styrofoam blocks to turn them into amphibious houses that can rise with floodwaters.

In addition, many both traditional and emerging ideas can help respond to sea-level rise, though these are beyond the scope of this book. While we will undoubtedly employ numerous engineering and/or green techniques to combat sea-level rise, we will also inevitably retreat from the rising sea. As the scale of the rise becomes clear, we will do what we can to save our major cities, including armoring the coast, that is, building hard structures to hold back the waters. But we must realize that in building hard structures on beaches, we will ultimately lose the beach (which might have been the very reason for building next to the ocean in the first place). Even though this may be an acceptable loss in order to save large cities, it is not acceptable on a large scale. Why protect the homes of the few (and largely wealthy) at the cost of losing the beaches that so many of us enjoy? Also, the cost of armoring the entire U.S. coastline would be astronomical. In the end, we will likely choose to defend our major cities and move back from most other areas.

Why move back? Why retreat? Why not hold the shoreline in place with massive seawalls or replenished beaches as the Dutch do? We can name three reasons:

1. As the sea level rises, the replenishment sand will become less stable, will erode faster, and will have to be replenished more frequently, and the cost will rise exponentially.
2. Seawalls built on eroding beaches will eventually cause the loss of the beach, and of course, the beach is why communities were established there originally.
3. In a few decades, the funding for holding the shorelines in place for tourist communities—such as those lining the beaches in Florida—will be overridden by the massive cost of saving cities such as Alexandria, Ho Chi Minh City, Amsterdam, Mumbai, Tokyo, Shanghai, Boston, Miami, and New York. When we have to spend billions of dollars to defend the big cities around the world, there will be little funding for the small municipalities and the tourist towns stretched along thousands of shorefront miles. This money drain for urban salvation will inevitably hasten the retreat from other shorelines.

Like it or not, we will retreat from most of the world's nonurban shorelines in the not very distant future. Our retreat options can be characterized as either difficult or catastrophic. We can plan now and retreat in a strategic and calculated fashion, or we can worry about it later and retreat in tactical disarray in response to devastating storms. In other words, we can walk away methodically, or we can flee in panic. These are the choices we have.

Like it or not, we will retreat from most of the world's nonurban shorelines in the not very distant future. Our retreat options can be characterized as either difficult or catastrophic.

2

The Overflowing Ocean

Larry Atkinson, a professor of oceanography at Old Dominion University in Norfolk, Virginia, pays close attention to the tide cycles. The route he takes to his office at the university normally takes him down Hampton Boulevard, but at every spring tide (when the moon is new or full), Atkinson takes Granby Street to work. The reason is that during the high tides, his usual route becomes flooded by saltwater forced up from the waters of Chesapeake Bay. If the wind is blowing onshore, the water level rises even higher. For the same high-tide reason, although the Unitarian Church of Norfolk is still holding services in its 112-year-old building, it now hopes to sell it. The problem is that flooding occasionally prevents access to the parking lot, so members of the congregation are urged to check the tide tables before coming to church.

Norfolk's original planners did not build a neighborhood that regularly went underwater at least twice a month. Rather, the tidal-flooding problem began only a few years ago. Sea level has risen to the point that spring tides routinely spill over into the fringes of the city. Normal high tides with just a minor onshore wind also penetrate the lower reaches of the city.

The city hired a consultant from a Dutch coastal-engineering firm to guide its plans to fight the sea with higher roads, better storm drains, and floodgates. The cost to Norfolk's population of 250,000 for engineering the shoreline and holding it in place will be about $1 billion. Even for that price, the city can be protected from only a 1-foot rise in sea level. Since 1920, the sea level at Norfolk has risen an extraordinary 1.5 feet. So far, the main response from those homeowners in Norfolk who can afford it has been to raise buildings 3 to 5 feet higher. New buildings are required to be elevated. Retreat has been mentioned, but only in passing. Norfolk has the same problem as Florida's high-rise–lined coast: the area around the city is flat and doesn't have an abundance of suitably elevated land on which to move buildings.

The How and Why of Sea-Level Rise

Norfolk, Virginia; Larry Atkinson's driving habits; and the problems facing the Unitarian Church of Norfolk all are harbingers of the future of coastal cities worldwide.

Wet Pavements

Norfolk isn't unique. All over the world, even on calm, windless days, waterfront roads, sidewalks, parking lots, buildings, drainage ditches, and docks are flooded by spring high tides. Local spring tide flooding is an expected and predictable phenomenon in parts of North America, from Queens, New York, to Miami Beach, Florida, to Galveston, Texas, to San Francisco, California, to Vancouver, British Columbia.

A 2014 report by the National Oceanic and Atmospheric Agency (NOAA) indicates that the rate of minor flooding, often referred to as *nuisance* or *tidal* flooding, causing at least wet pavement or ponded parking lots, has increased significantly in recent years. Before 1971, Wilmington, North Carolina; Washington, D.C.; Annapolis, Maryland; Atlantic City, New Jersey; Sandy Hook, New Jersey; and Charleston, South Carolina, all experienced, on average, fewer than five flood levels per year. But since 2001, these communities have undergone an average of more than 20 each year. Annapolis seems to have suffered the most. In the 1950s, the number of nuisance floods there averaged five per year, but now the number is close to 40. The flooding is intensified if wind is blowing onshore or if the high tide is a *king tide*, an especially high spring tide that occurs two or three times a year when the moon and the sun are oriented just right to create a larger than usual tide range.

The Union of Concerned Scientists predicts that by 2045, the fringes of Washington, D.C.; Annapolis, Maryland; and Wilmington, North Carolina, will have nuisance flooding every day of the year. By 2045, Atlantic City, Miami, and Baltimore will suffer tidal flooding about 250 days each year.

King Tides

King tide is a colloquial term for the very highest of tides. In New South Wales, Australia, the Department of the Environment came up with the brilliant idea to use king tides to educate the public about sea-level rise. This led to the Witness King Tides community photography project, which has spread, as the King Tide Photo Initiative (KTPI), to a number of U.S. states, British Columbia, and even Tuvalu, an island nation in the Pacific. Each of these organizations announces to the public the dates of the next king tide and urges citizens to document the now-routine flooding with photos to display online. These photographs of king tides offer a glimpse into our future—a preview of where the increasing levels of the sea are headed and an engaging way to alert and educate the concerned, the skeptical, and the unaware.

The Mechanics of Sea-Level Rise

The consensus among climatologists, glaciologists, and oceanographers is that our society should be prepared for a 3-foot rise by 2100. For Norfolk, Virginia, where sea-level rise is unusually rapid, the expected rise could be 5.5 feet. Norfolk has such a high

The consensus among climatologists, glaciologists, and oceanographers is that our society should be prepared for a 3-foot rise by 2100.

rate of sea-level rise because it is simultaneously affected by four of the main causes of sea-level change: the increased volume of seawater from melting ice; the increased volume of seawater from warming oceans; the subsiding land; and the changes in ocean circulation causing the central Atlantic Ocean bulge to flatten and contribute water to the mid-Atlantic coast. We discuss these causes later.

Estimates of future sea-level rise are largely based on mathematical models, which in turn are based on an escalating series of assumptions, most of them good but some a bit fuzzy. Climate model uncertainties (the fuzzy bits) include

- Rate of production of CO_2 in a future industrial world
- Future cloud cover as the earth warms (i.e., low-lying white clouds reflect back some of the sun's radiation, a process that would reduce the rate of atmospheric warming, whereas higher cirrus clouds absorb radiation, reinforcing the greenhouse effect)
- Local changes in atmospheric processes
- Climate variability
- Future behavior of the ice sheets in Greenland and Antarctica

The real world, with melting ice, failing ice sheets, warming ocean water, and both tide gauge and satellite measurements, leaves no room for doubt as to the future of our shorelines. We, the people, are the primary cause of sea-level rise because of our production of greenhouse gases, particularly carbon dioxide, which prevent some of the sun's heat from reflecting back into space. As a consequence, the heat stays in our atmosphere and thereby warms the earth. The warmed atmosphere then causes the ocean waters to expand and the ice sheets to melt, leading to a rising sea level.

We, the people, are the primary cause of sea-level rise because of our production of greenhouse gases, particularly carbon dioxide, which prevent some of the sun's heat from reflecting back into space.

Measurement of the Sea Level

Sea-level changes are measured from the geoid in two ways, by means of either *tide gauges* or *satellites*. The *geoid* is the theoretical surface of the ocean, assuming no winds or currents. All told, there are more than 1,700 long-term tide-gauge records worldwide, most concentrated in the Northern Hemisphere. The oldest records go back about 150 years, although an Amsterdam tide record goes back to the year 1700. But because most gauges are situated in sheltered water rather than in the open ocean, they may produce apparent but not real changes in sea level. This could happen when a new channel or inlet is dredged or when an inlet closes up, affecting the volume of water flowing into and out of a sheltered bay. Instead, the ideal location for a tide gauge is on some sort of stable platform (like a concrete pier) along an open ocean shoreline.

Tide gauges measure *relative sea level*, the level of the sea with respect to the location of the gauge. Thus, if the land is moving up, as in those parts of Scandinavia where the weight of glaciers has been removed, the tide gauge may show a drop in sea level. On deltas, the land is often sinking, providing a tide record that indicates a sea-level rise more rapid than the global rate. In both these examples, the gauge is measuring the combined movement of the earth's surface and the ocean's surface.

Over the last 20 years, satellite altimeter measurements have changed. Now we have millions of measurements of the sea surface in the open ocean and have gained a startlingly detailed understanding of the global variations in sea level. Maps showing sea-level change all over the globe indicate the strong effects of winds, ocean currents, rainy seasons, and El Niño, a band of unusually warm water in the equatorial Pacific with a variety of side effects. The current global sea-level rise calculated by using satellite data is 0.1 inch per year, or around 1 foot per century.

A third kind of sea-level-change data, on a much longer timescale, that may help us understand the future, is the evidence of rates of past sea-level changes preserved in the "recent" geologic record. This could provide a reality check for what sea-level rise may have in store for us. For the last 2 million years, the earth has undergone at least eight major ice ages during which sea level has gone up and down as the glaciers released or captured water. Thus, although the sea level has always varied widely, the difference now is that we have built cities, towns, and infrastructure on the shore that are directly affected by the rising seas.

Ancient sea-level changes are measured by mapping preserved deposits from past shorelines. This includes those that have been submerged and overrun by a rising sea as well as those stranded above the current shoreline, left behind when a new ice advance came along and the sea level dropped. Radiometric dating of submerged ancient coral heads that once flourished at sea level, as well as beach sand deposits well above and far inland from today's shoreline, can provide clues to or indicators of how fast the sea level is moving up and down.

Although the sea level has always varied widely, the difference now is that we have built cities, towns, and infrastructure on the shore that are directly affected by the rising seas.

The Causes of Sea-Level Change

Global Changes

The sea level rises and falls for a number of reasons. The global cause with which we are most concerned, because it operates on a timescale of interest to humans, is the change in the volume of the sea. This rise comes mainly from the melting of the world's ice sheets and mountain glaciers and from *thermal expansion*, the increase in the ocean's volume as it absorbs heat from the atmosphere. During most of the twentieth century, the thermal expansion of the water in the upper 2,000 feet of the ocean was the main cause of sea-level rise, but now it accounts for about half. In the twenty-first century, meltwater from ice will likely take over as the driving force behind sea-level rise.

The potential for sea-level rise from the world's ice bodies goes something like this: If all of Greenland's ice melted, the sea level would rise 20 feet. If the West Antarctic Ice Sheet melted completely, the seas would rise 16 feet. The much larger East Antarctic Ice Sheet would raise the seas by 164 feet, and all the world's mountain glaciers would contribute between 1 and 2 feet.

The most obvious and compelling evidence of global climate change is the retreat of mountain or outlet glaciers. An *outlet glacier* is one that flows from a less mobile, central ice cap. Outlet glaciers are found extending from both the Greenland and Antarctica Ice Sheets as well as in Iceland, Canada, the Himalayas, and in Patagonia in Argentina and Chile.

The Jacobshavn Glacier in western Greenland produces 10 percent of all of Greenland's icebergs and was responsible for an estimated 4 percent of the global sea-level rise in the twentieth century. This area produced the iceberg that sank the ship *Titanic* in 1912, which was memorialized in 1997 in *Titanic*, a blockbuster film directed by James Cameron. In recent years, Cameron has become an environmental advocate who speaks out on global warming and sea-level rise.

In the past few years, the movement of the Jacobshavn Glacier to the sea has sped up to 150 feet per day, becoming the fastest-moving major glacier in the world. The glacier had been jammed up by a bump on the seafloor, but the thinning of the ice from below suddenly released the glacier from the bump and led to the spectacular speedup. Since 1500, the Mendenhall Glacier near Juneau, Alaska, has retreated 2.5 miles, with 1.75 miles of that retreat occurring since 1958 when a lake formed in front of the glacier. A glacier terminating in a water body, whether in a lake or the sea, calves and loses ice at a faster rate than does a glacier that is entirely on land. As the Mendenhall Glacier retreated, the land *rebounded* from the loss of the weight of the ice, thus causing the relative sea level to drop along Juneau's shoreline.

Global warming is not the only condition that affects the rate of glacier retreat. In a 2013 study, National Aeronautics and Space Administration (NASA) scientists found that between 1860 and 1930, a time when local atmospheric temperatures dropped slightly, large valley glaciers in the Alps nevertheless retreated by 0.66 mile. This was a rate of retreat that had not been seen in recorded history. Why did this happen, when the atmosphere wasn't warming? The study of ice cores from those glaciers showed that very likely the anomalous retreat was caused by black carbon or soot from industry that reduced the reflectivity of the ice and thus increased the glaciers' absorption of heat. The villain was the Industrial Revolution! Even today, industrial soot and volcanic ash continue to affect melting rates around the world.

Satellite measurements are as important on ice as they are over the ocean. *Satellite altimetry* measures the elevation of the ice sheet. *Satellite gravity* measurements on Greenland and western Antarctica determine the mass of the ice and answer the important question of whether there is an annual net gain or loss of ice. The answer is that there is a significant loss of mass (thinning) in both Greenland and western Antarctica, which indicates that the winter snowfall does not balance out the loss of ice during the summer melt season.

The Greenland Ice Sheet's rate of loss is controlled primarily by global warming and the slope of the land. The Antarctic Ice Sheets' loss of mass is more complex than that of the Greenland Ice Sheet. Because of the extremely low temperatures, very little ice melts on Antarctica. The huge East Antarctic Ice Sheet lies almost entirely on land, largely ringed by ice shelves, and so far is shedding very small amounts of ice into the sea. But western Antarctica has a number of outlet glaciers that are grounded on offshore islands or bumps on the continental shelf, and as these glaciers thin, they will become detached or ungrounded, causing them to flow seaward and accelerate. This will set the scene for a rapid loss of ice (and a rapid increase of water flowing into the sea).

The Greenland Ice Sheet's rate of loss is controlled primarily by global warming and the slope of the land. The Antarctic Ice Sheets' loss of mass is more complex than that of the Greenland Ice Sheet.

In addition, the warming seawater along the western margin of the Antarctic continent is a major cause of the recent uptick in the rate of ice melting. The seawater warms as newly formed wind patterns blow the colder near-surface seawater offshore, which then is replaced by warmer water from below the sea surface. Normally, the deeper ocean water is cooler than the surface water, but along Antarctica, the reverse is true because of the melting sea ice in summer and icebergs calving from the ice sheets.

A factor of great importance contributing to the future sea level is the *ice shelves.* These are flat, floating ice platforms, 300 to 3,000 feet thick, positioned around the seaward margins of ice sheets, particularly in Antarctica. Two very large ones around the margin of western Antarctica are the Ross Ice Shelf (166,000 square miles) and the Ronne-Filchner Ice Shelf (188,000 square miles). Most of the large outlet glaciers abut these shelves, which are thinning from below as the oceans warm, and already they are making a significant contribution to the rising sea level. As the shelves weaken and eventually break up, the glaciers will suddenly be unobstructed and will flow into the sea, where they, like the ungrounded glaciers, will release more meltwater. Studies reported in 2014 indicate that the West Antarctic ice shelves are thinning and may now be in the process of breaking up. In 2014, NASA glaciologist Eric Rignot coauthored a paper and presented the results at a conference arguing that the deterioration of the Amundsen Sea portion of the West Antarctic Ice Sheet (an area almost the size of France) was now unstoppable. The collapse of this portion of the ice sheet, perhaps over a time period of two centuries, will, by itself, raise sea levels by 3 feet. *Mother Jones* magazine characterized this observation as a "holy shit moment" for global climate change.

Several relatively small ice shelves, such as the Larsen B on the Antarctic peninsula, already have broken up. Scientists believe that the large, sudden collapse of the Larsen B Ice Shelf in 2002 may have been caused by the ice rebounding from the loss of the weight of thousands of small meltwater lakes on the surface that suddenly drained through fractures in the ice during a (relatively) warm summer. This caused a chain reaction in which stress fractures propagated to the base of the ice, leading to the rapid disintegration of the Larsen B Ice Shelf, which was approximately the size of Rhode Island. The collapse of the last remnant of Larsen "is most likely in progress," according to California scientists.

Local Changes

Local sea-level changes are sometimes of overriding importance in affecting coastal communities. River deltas, including the Mississippi, the Ganges, the Niger, the Yangtze, and the Nile deltas, are home to more than 350 million people and are *subsiding* (sinking) because of compaction of the large sediment load that accumulates at river mouths over thousands of years. In addition, dams upstream from most deltas have cut off the supply of fresh river sediment. Although this sediment would make up for some of the sinking, other factors are evident as well. For instance, oil and water are being extracted from some deltas, which adds to the sinking and, of course, increases the local sea-level rise in those areas. One example is a portion of the Mississippi Delta, where the sea level is rising at a rate of 4 feet per century, mostly as a result of subsidence, in contrast to the overall global rate of 1 to 1.5 feet per century.

River deltas, including the Mississippi, the Ganges, the Niger, the Yangtze, and the Nile deltas, are home to more than 350 million people and are *subsiding* (sinking) because of compaction of the large sediment load that accumulates at river mouths over thousands of years.

Along some mountainous coasts, local sea levels change because of *tectonic forces* that are part of the mountain-building processes. For example, the Pacific coast of Colombia at the foot of the northern Andes has frequent small earthquakes that cause local shorelines to drop as much as 3 feet overnight (resulting in an overnight rise in sea level of 3 feet) along shoreline distances of 50 miles or so. Shoreline erosion rates increase immediately, and local people move their houses, which are designed to be easily moved away from the shore. Overall rates of sea-level rise may be as high as 10 feet per century along the Pacific coast of Colombia. Earthquakes can work both ways, however. The Great Alaska Earthquake of 1964 affected a huge area and both lowered and raised local sea levels.

Changes in *ocean circulation* are responsible for much of the local sea-level change in all of the world's oceans. In recent years, the sea level has risen two to three times faster than the global rate along the U.S. Atlantic coast from Cape Hatteras, North Carolina, to Cape Cod, Massachusetts, because of a slowing down of the Gulf Stream. This narrow, fast-moving current started to slow down in 2004, which lowered the elevation of the central Atlantic Ocean, which had been "held up" by the forces of the Gulf Stream. This lowering of the central ocean bulge pushed water toward the Atlantic coastal waters, accelerating the local sea-level rise in the process (and adding to Larry Atkinson's driving-route problem in Norfolk, Virginia). In 2014, University of Arizona and NOAA scientists reported in an article in *Nature Communications* that the sudden surge of almost 4 inches in sea-level rise in 2009/2010 between New York and Canada was probably due to a slowdown in a major ocean current system known as the Atlantic Meridional Overturning Circulation (AMOC), of which the Gulf Stream is a component.

The *gravitational pull* of large masses of ice, as on the Antarctic continent, pulls the ocean toward the continent and raises the local sea level there. As the ice melts, the local nearshore sea level slowly drops because the gravitational pull of the ice mass decreases (and the land rebounds from the loss of the weight of the ice). Of course, though, the melting ice causes the sea level to rise elsewhere.

The *cycle of water* on dry land leads to both the rise and the fall of the sea level. The widespread clearing of land for agriculture reduces the absorption of water by forests and vegetated soils and causes more runoff to the sea, as does the vast increase of impervious surfaces in cities and towns. Likewise, the extraction of groundwater at rates higher than the amount of compensating rainfall also adds water to the sea.

The widespread clearing of land for agriculture reduces the absorption of water by forests and vegetated soils and causes more runoff to the sea, as does the vast increase of impervious surfaces in cities and towns.

Conversely, both trapping water in reservoirs behind dams and irrigating arid lands, where much of the water soaks into the groundwater, decrease runoff to the seas. Overall, the portions of the water cycle on land that cause the sea level to rise and fall more or less cancel each other out.

The graphs of sea-level rise over time are what scientists call *noisy* because of the year-to-year and decades-long variations in some of the factors just cited. Because of the many peaks and valleys on the line marking the sea level, at least a decade of data is required in order to sort out any trends. One downward trend lasted for 18 months from 2010 into early 2011 when the global sea level dropped about 0.3 inch. According to a study published in *Geophysical Research Letters* in 2013, the probable cause of this temporary drop was the number of torrential rainfall events in over a 2-year period (2010/2011) in Australia, coupled with the dry soils of the Australian desert and its poorly developed drainage system. Much of the rain did not immediately reach the sea. Since then, however, the rate of sea-level rise has resumed its pre-2010/2011 pace.

A Voice in Unison: The IPCC

The United Nations' Intergovernmental Panel on Climate Change (IPCC) may mark the first time in global history that the world's scientists have joined together to warn the world about a perceived impending natural disaster. The IPCC has issued five IPCC reports, in 1990, 1995, 2001, 2007, and, the latest, 2013. The IPCC has an annual budget of around $9 million and an administrative staff of only 12 people who work out of a small office in the World Meteorological Organization building in Geneva, Switzerland. A budget of only $9 million seems like only a tiny drop in the bucket for an organization of such importance!

The actual work of the IPCC is carried out by thousands of scientists who work for no salary. Not surprisingly, coming up with something even close to a consensus in these large groups is taxing and time-consuming in the extreme. Nonetheless, the fact that committees consisting of hundreds of individuals can come up with any report at all is admirable. Perhaps the most important document from the IPCC is the Summary for Policy Makers (SPM), a report prepared by a dozen or so experts who draw conclusions from the more comprehensive documents. Then comes the kicker. The representatives of every government in the world that wishes to join the IPCC cause must approve the SPM document or else drop out. Long, intense meetings are held at which important issues as well as adjectives, commas, and semicolons are discussed in order to come out with the final SPM. The very existence of the SPM, therefore, seems like a miracle! No wonder it is often densely written.

> The actual work of the IPCC is carried out by thousands of scientists who work for no salary.

After the 2007 IPCC report, issues arose that affected the whole global climate-change debate. The first was the disclosure of e-mails involving IPCC writers allegedly trying to suppress data. This turned out to be grousing among colleagues, blown out of proportion by those who sought to deny the existence of climate change. A second problem was the inclusion of a mistaken non-peer-reviewed estimate in the report noting that the Himalayan glaciers would disappear with startling rapidity in 35 years, a highly unlikely rate of loss.

Because there is an army of global-change deniers waiting to pounce on the slightest perceived inaccuracy, the IPCC tends, if anything, to be too conservative. For example, so far the panel has consistently been low in its estimate of sea-level rise rates. The 2007 IPCC report did not include ice-sheet melting in its projections, feeling that more information about measured rates of melting was needed, but the 2013 report does include ice sheet behavior in the estimate of sea-level rise. Another bit of conservatism involved the Arctic Ocean's sea ice. Although the 2007 report suggested the summer sea ice would be gone by 2100, it now appears that in a worst-case scenario, it more than likely could be gone by 2020.

The IPCC's 2013 report emphasizes a couple of issues: One is the apparent slowing down of warming during the last 15 years (very recent reports argue that there wasn't a slowdown). The earth is still warming, but at a slower rate as atmospheric CO_2 continues its upward march to 400 parts per million (ppm) and beyond. The slowdown then brings up whether the climate is less sensitive to carbon dioxide than we've been assuming. Scientists believe that one explanation is that the oceans are absorbing much of the increased heat. But more fundamentally, it is well established that the increasing heat in the atmosphere is determined by more than just its CO_2. There are variations in the rates of release of other greenhouse gases, such as methane from the melting permafrost. And there are natural variations related to the amount of solar radiation that reaches the earth. Thus, occasional slowdowns and speedups are to be expected.

How Fast and How Far

Currently, a widely accepted sea-level rise scenario is a minimum of 1.5 feet, a maximum of 5 feet, and a likely rise of 3 feet by the year 2100.

In the 1980s, James Hansen, former director of NASA's Goddard Institute for Space Studies, was the first to bring global climate change and sea-level rise to the attention of the U.S. Congress. Then in 2015, Hansen and 16 colleagues dropped a bombshell prediction of a possible maximum 10-foot sea-level rise in this century. This number was based on the assumption that once the ice sheets begin to break up, melting will proceed at an accelerating, rather than a steady, pace.

In another paper, published in 2013, Hansen and three others argued that "burning all fossil fuels would make most of the planet uninhabitable by humans, thus calling into question strategies that emphasize adaptation to climate change." In other words, we can't neglect the critical importance of reducing atmospheric carbon dioxide. Climate scientist Ken Caldeira, coauthor of a 2015 paper that looks at what could happen if we burned up all available fossil fuels, predicts a sea-level rise on the order of 100 feet in 1,000 years. With a sea-level rise of that magnitude, we would be forced to abandon coastal cities worldwide.

Other scientists approach the problem of predicting the future rise of the sea by studying the geologic past. Because we have been able to observe the behavior of the sea level for only (geologically speaking) a very short time, we must go to the geologic past to expand our horizon and get a peek into what might be in store for us.

The Quaternary period (which includes the Pleistocene [ice age] epoch) represents the last 2.6 million years of geologic time. During the Quaternary, massive glaciers often advanced over large areas of land far from the poles. Between each major glacial advance was a warmer interglacial period when the climate was not much different from what it is today. Now we are in an interglacial time known as the Holocene and the Anthropocene (a newly recognized segment of geologic time that indicates when we humans began to have a major impact on the earth's surface, after the onset of the Industrial Revolution). The previous interglacial, the Eemian interglacial period, is assumed to have begun 130,000 years ago and to have ended about 114,000 years ago when the glaciers began to accumulate ice once again. The Eemian period (also referred to as Marine Isotope Stage 5e) is significant today because the events of that time come the closest as an example of where the current sea-level rise may be headed. With each of the massive glaciers' advances and retreats, the level of the sea changed dramatically. When water was tied up in the ice, the sea was much shallower, and as the ice melted, the sea rose.

One of the best and most comprehensive discussions of sea levels past, present, and future is *Rising Seas*, by Vivien Gornitz. Geologists like Gornitz understand quite well the big picture of sea-level rise on a scale of tens of thousands of years. The fact that the sea level has risen 400 feet in the past 18,000 years and came close to its current level 5,000 years ago is easily demonstrated in the geologic record. But understanding how fast and how far sea levels in the past have risen over short periods of time, such as a century or two, is a different matter. It is the potential for short-term changes that is of the most interest to our society today. We must keep in mind that it is human activity, mainly the burning of fossil fuels, which was not a factor in the geologic past, that is driving the current sea-level rise. The IPCC's 2013 report expresses 95 percent confidence that we humans are the problem, that we are producing the CO_2 and other greenhouse gases that are the cause of global climate change.

The sea level during the Eemian interglacial was above the current shoreline, probably hovering around 20 feet higher for most of the time. The more recent sea-level rise events, as recorded in beach or salt marsh deposits or coral heads stranded on land far from the sea, provide an important indication of how the sea level changed in the past. Gornitz's global summary of field evidence of ancient shorelines shows that at times between 121,000 and 119,000 years ago during the Eemian, the sea level rose suddenly by 20 feet or more.

In the same time frame, very rapid changes in sea level (as much as 8 feet per century) are suggested by oxygen isotope studies of Red Sea foraminifera, tiny organisms that live on the seafloor. The oxygen isotope composition of foraminifera in long marine cores reflects the volume of ocean water locked up in ice, which in turn translates roughly to sea level.

In 2013, Australian Michael O'Leary and five colleagues published a study of the Eemian shoreline evidence along 1,000 miles of the coast of western Australia. They chose this location because Australia's west coast has not experienced geologically recent ups and downs caused by the weight of glaciers or by deep-seated crustal forces. Because the earth's movements have not distorted the sea-level change record there, the field evidence collected from the west coast is assumed to be particularly strong. These scientists found evidence of a jump in sea level of 17 feet during the last interglacial that occurred over a maximum period of 1,000 years (but possibly a shorter time within the 1,000-year span). The accuracy of their dating techniques would not allow a more exact estimate of the timing of the leap. Nonetheless, the important takeaway from this research is that a significant sea-level rise can be rapid, and thus we (i.e., present-day civilization) must recognize that possibility and prepare accordingly.

The world is already experiencing and is locked into a warmer climate and rising seas, no matter how successful we are at reducing CO_2. This basically involves the momentum of change. Anders Levermann, the lead author of the latest IPCC report's chapter on sea level, argues that in order to halt a rise, atmospheric CO_2 must actually be removed from the atmosphere, not just reduced in the quantity produced by humans. He compares the rise in sea level to a big ball: “It takes a while until you get it rolling, but once it’s rolling you can’t stop it easily.” Carbon dioxide has an atmospheric lifetime of thousands of years, and as a consequence, the CO_2 now held in the atmosphere will keep air temperatures warm even if CO_2 production is stabilized or reduced.

Carbon dioxide has an atmospheric lifetime of thousands of years, and as a consequence, the CO_2 now held in the atmosphere will keep air temperatures warm even if CO_2 production is stabilized or reduced.

Levermann’s point is that we have enough heat in the atmosphere to do lots of damage right now. Sea-level rise will continue for a long time even if atmospheric warming doesn’t continue to increase but instead remains steady. The reason for this long lag-time for reversal is that the oceans and the ice masses are very large, and atmospheric warming penetrates “slowly but inevitably.” The result is that sea levels will continue rising for centuries and perhaps millennia.

Levermann and his co-workers believe that over the next 2,000 years, we can expect 7.5 feet of sea-level rise per degree Celsius (1 degree Fahrenheit should raise the sea level about 4 feet) of atmospheric warming as the melting of the world’s big ice sheets, which they refer to as “slumbering giants,” overtakes the ocean’s thermal expansion as the principal cause of sea-level rise. How long the sea level’s response (to each degree of warming) will take is unknown. Certain that the sea level will take some time to “catch up,” Levermann and his colleagues believe that the momentum of change in the huge masses of the oceans and ice sheets “imposes the need for fundamental adaptation strategies on multicentennial time scales.” In other words, we need to be planning on a long-term basis.

Andrea Dutton and several coauthors of a 2015 paper looked at five different warm periods in the geologic past and found that at least 20 feet of sea-level rise had occurred in each instance. Their result agreed with Levermann’s. A sea-level rise of 15 or more feet is a possibility with 2 degrees Celsius of warming of the atmosphere. No time frame is predicted for such a rise.

The Tragedy of the Bruun Rule

The big remaining question is how far and how fast shorelines will retreat in response to a given rise in the sea. On one scale, the answer is obvious. The shoreline will retreat faster on a gently sloping coast than on a steep mountainous coast. The same goes for a sandy coast versus a rocky coast. But these generalizations are not usually of sufficient accuracy to help a coastal property owner or an entire community trying to understand their future in a rising sea.

Predicting shoreline retreat is an immensely complicated process. Storms are a major direct cause of retreat, and they are impossible to quantify because of their random occurrence and variability in winds, direction, and duration. No one has come up with a successful way to predict future shoreline retreat, and probably no one ever will. Understanding shoreline retreat requires an understanding of continental shelf processes, beach processes, future wave conditions including storm waves, a detailed knowledge of the geology of the land, plus an understanding of how the shoreline has responded to past storms, large and small. Probably the most widespread method of predicting future shoreline retreat rates is by projecting the current annual erosion rates into the future.

One common answer to the big question all over the globe is that for every 1 foot of sea-level rise, there will be 100 or 200 feet of shoreline retreat. This absurd belief, based on a simple mathematical model called the Bruun Rule, is a vivid example of the problems with blindly accepting mathematical models. The rule, named for its inventor Per Bruun, a Danish American coastal engineer, is popular because it is easy to use and needs only two pieces of information to apply. One is the local rate of sea-level rise and the other is the slope of the local shoreface or inner continental shelf. The latter number can be obtained from any reasonable navigation chart.

The Bruun Rule makes some invalid assumptions, which include these three:

1. It does not take into account the slope of the land surface over which sea-level rise will push the shoreline.

2. It assumes without any basis that the slope of the shoreface is the only control of shoreline retreat besides rising seas.
3. The model makes a number of other implicit assumptions that are not met in nature, including storm-related processes.

Perhaps the most important reason that the rule (or its tweaked modifications) has survived is that no other model has taken on the task of solving such a hugely complex question. The Bruun Rule is a classic case in which a mathematical model has done real damage to society by underplaying the potential of a rising sea to push back the shoreline. In fact, the slope of the land surface over which the shoreline or barrier island will move in response to a rising sea must be a major factor. In North Carolina, the lowermost coastal plain slopes average 1 to 2,000, meaning that in theory, a 1-foot rise in sea level could move the shoreline back 2,000 feet. On the Outer Banks of North Carolina, the regional slope of the land is 1 to 10,000, which means in theory that a 1-foot rise in sea level could move the shoreline about 2 miles.

A rise of 3 feet would be a disaster for coastal populations worldwide. The focus has been on the projected 3-foot sea-level rise that is anticipated during this century, but the seas are expected to continue to rise for centuries afterward. Coastal civilization won't end in 2100, so there is nothing magical about choosing that date as a planning goal. In fact, coastal planning should be looking far ahead, as the Dutch do.

3

The Fate of Two Doomed Cities

Miami and New Orleans

Two vibrant American cities squarely in the cross-hairs of sea-level rise are Miami and New Orleans. What makes Miami exceptionally vulnerable is its unique geology—that is, the nature of the city's underlying rock. What makes New Orleans vulnerable are its low elevation on a subsiding delta and its dependence on dikes. There is no doubt of the future for these two troubled global cities; both cities as we know them are doomed. The only question is how they should respond to sea-level changes before their demise.

Miami: A City with Its Head in the Sand

By almost any measure, across the world the city most threatened by sea-level rise is Miami, Florida. Miami proper has a population of 400,000, but when you consider the entire metropolitan area, the population is closer to 5.5 million. The average elevation of this fourth largest urban area in the United States is just 6 feet above sea level, and its highest natural feature is a ridge that averages only 12 feet. Large areas of the city are a mere 2 to 3 feet or less above sea level. With sea-level rise, Miami also faces major challenges in flood control as well as contamination of its drinking water by salt (salinization). Related problems include storm-sewer backflow and overflow, and water-treatment systems that will be flooded. When the water rises

2 feet, the county's sewage treatment plant and the Turkey Point Nuclear Generating Station will be islands in the sea (discussed further in chapter 8).

During spring tides, full- and new-moon days when tides are at monthly peaks, those who cross on the MacArthur Causeway from mainland Miami to the Miami Beach barrier island may find themselves hubcap deep in saltwater when they drive off the causeway onto Fifth Street and Alton Road on South Beach. Those who stroll along North Bayshore Drive in Miami may find their shoes and socks wet from the spring tide. The situation worsens when heavy rains coincide with the spring tide because there is nowhere for the freshwater to drain, as seawater has reversed the flow of the storm drains. The Union of Concerned Scientists believes that within two to three decades, Miami Beach may be flooded 237 times per year. Rarely mentioned is the fact that heavy rains are accompanied by floodwaters that give off a strong odor of human waste originating from backed-up sewage.

Aside from catastrophic floods during hurricanes, the worst flooding occurs when onshore winds coincide with heavy rains and spring tides. Similar flooding events are increasingly common along Mola Avenue in Fort Lauderdale to the north and in the Florida Keys at the southwestern tip of the state.

Inundation

Porous limestone underlies most of South Florida, including Dade County and greater Miami. The Miami Limestone formation, sometimes called the Miami Oolite, is the most permeable and porous formation (figure 1). This 40-foot-thick limestone layer consists largely of oolites and bryozoan skeletons that were deposited in shallow water about 125,000 years ago during the Pleistocene ice age. *Oolites* are small spherical sand grains made of calcium carbonate that precipitated out of an ancient warm shallow sea, like the oolites found today on the Bahama Banks. *Bryozoans* are the skeletons of tiny plant-like animals (marine invertebrates), which also are composed of calcium carbonate.

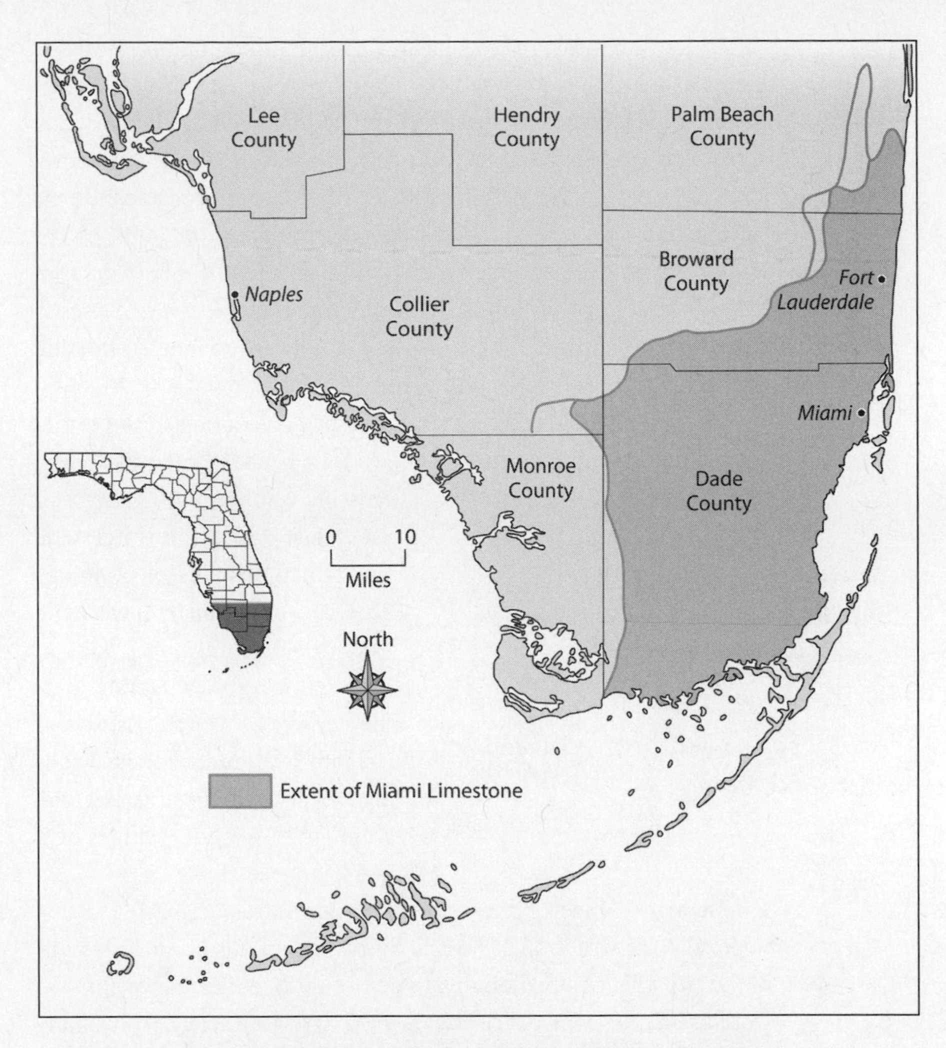

FIGURE 1 A sketch map showing the aerial extent of the 125,000-year-old Miami Limestone in southeastern Florida. This limestone is very porous and permeable and underlies most of the greater Miami area. Because of the limestone, rising sea waters will flow into the city whether or not it is surrounded by seawalls. (Courtesy of the Florida Department of Environmental Protection's Florida Geological Survey)

Porosity is a measure of the pore space in a rock, and *permeability* is a measure of the rate at which a fluid like water can move through the interconnected pore spaces. If you picture a sponge made out of stone, you can begin to understand the porosity of the Miami Limestone. The most spectacular evidence for its extreme permeability is the tidal fluctuation seen in some freshwater ponds within the city of Miami. Careful measurements of pond levels reveal minor fluctuations that correspond closely to the offshore tides. As the tide rises at the beach, the levels of the ponds rise a fraction of an inch. As the tide goes out, the pond levels are lowered.

What this means is that Miami, and much of southeastern Florida extending two-thirds of the way up to Fort Lauderdale, is like a city on stilts whose buildings are elevated just above the water. The siting of thousands of buildings here can be compared with the Neolithic dwellings on pilings in Lake Zurich in Switzerland, or the "crannogs" of Scotland and Ireland, or the modern lake dwellings of Myanmar and Nigeria, although the city of Miami is on a much larger scale.

What this means is that Miami, and much of southeastern Florida extending two-thirds of the way up to Fort Lauderdale, is like a city on stilts whose buildings are elevated just above the water.

The significance of the porous limestone to Miami's future is huge. The construction of levees, dikes, or seawalls around all sides of the city will not stop flooding by sea-level rise. Although such walls would reduce the impact of storm surge and wave attack from a hurricane, they would not reduce inundation from a rising sea. Lowering the water level within a levee-enclosed Miami with giant pumps like those in New Orleans would be an exercise in futility. Pumping out the water flooding the city from all sides as a result of sea-level rise (assuming the city was completely surrounded by a levee or dike) would be like attempting to lower the level of the ocean with pumps. As fast as the water could be pumped out, the same volume from the sea would come in.

At risk from inundation in Miami-Dade County are billions of dollars in residential and commercial real estate, plus hundreds of schools, hospitals, sewage plants, power plants (including two nuclear reactors), farmlands, landfills, and hundreds of hazardous material sites.

Drainage-Canal Floodgates

Much of the development of Miami and Dade County was carried out by converting Everglades land (draining the swamps) into land suitable for construction. The resulting flat and low-elevation ground, however, was frequently flooded from rainfall for long periods of time. To combat this, the state built a canal system to drain the floodwaters into the Everglades or the ocean. Eventually more than 800 miles of canals were constructed, 1 to 2 miles apart. An immediate bit of collateral damage from canal construction was the intrusion of saltwater into the Biscayne Aquifer (the body of groundwater that supplies freshwater to South Florida) as storm surges pushed seawater far inland. To halt salinization of the community's freshwater source, flood control gates were needed. The idea was that when a storm threatened, the gates would be closed temporarily to prevent the intrusion of saltwater into the interior of the city. Closing the flood control gates during storms did reduce intrusion, but unfortunately it also prevented the canals from draining rainwater and led to the flooding of nearby neighborhoods.

Most of the construction of the floodgates took place 50 to 60 years ago. Since that time, Florida's local sea level has risen as much as 8 inches. Already a number of the gates are unable to discharge rain runoff when the tide is high, and as the sea level rises, more of the gates will become useless. Replacing the gates would be an ongoing effort costing millions of dollars. Eventually, of course, draining South Florida via canals will be impossible once the seas inundate much of the land. So in the near future, freshwater flooding during rain events will become increasingly the norm in South Florida.

The Salinization of Drinking Water

As the sea level rises, destruction of the freshwater supply through salinization and overuse becomes an increasing problem. As a result, the groundwater along all the ocean rims of the state has already been contaminated by saltwater. There are several different approaches to resolving this problem. The first and cheapest approach is simply to move the wells to the west, away from the ocean. Of course, this would be a temporary move because eventually the saltwater will penetrate to the new wells.

A proposed multibillion-dollar Everglades restoration project would include filling in canals and restoring river meanders. If this were done, the amount of available freshwater would temporarily increase. It is doubtful, however, that the proposed project could be successful in a time of rising sea level and, for this reason, may well be abandoned.

Another alternative is to desalinize the contaminated water in the Biscayne Aquifer or to desalinize the much saltier ocean water (not a likely alternative in view of the cost of desalinization). In either of the two remedies, the resulting brine by-product would be pumped into deep wells for "permanent" safe storage.

The Final Straw: Political Ineptitude

It is a tragedy that a state so clearly threatened by sea-level rise is guided by politicians so blind to the dreadful future their own people face. Developers who routinely continue to build waterfront skyscrapers share the blame.

In the official Republican response to President Barack Obama's 2013 State of the Union address, Marco Rubio, a U.S. senator from Florida and presidential candidate, dismissed the government's response to sea-level rise. Rubio, appointed to chair the Subcommittee on Oceans, Atmosphere, Fisheries and Coast Guard in 2015, stressed that "our government can't control the weather." He also declared that since he is not a scientist, he is not qualified to have an opinion about the human role in climate change. Both Rubio and

Florida's governor, Rick Scott, are skeptical about the human connection to climate change and sea-level rise. During the gubernatorial campaign, Scott asserted that he did not believe in climate change, despite having been briefed by a panel of climate scientists about future sea levels. Another Floridian who is also a presidential candidate, former governor Jeb Bush, is in the same camp, describing himself in a July 2009 *Esquire* magazine article as a skeptic and reminding us all that he is "not a scientist." Thus, Florida's response to climate change is complicated by the current political climate and lack of leadership and even its politicians' polltical cowardice.

Florida's response to climate change is complicated by the current political climate and lack of leadership and even its politicians' political cowardice.

Ironically, when Scott's predecessor, Charlie Crist, was the governor, sea-level rise already was an important issue. In a speech before the state legislature, Crist announced that global warming was one of the most important issues that we would face in this century. But times change, and with the election of Scott, instead of leading on the issue, the Florida state government has swung back into the denial camp. According to Tristram Korten, in a March 2015 report from the Florida Center for Investigative Reporting, employees of the state environmental agency were told not to use the terms *climate change* and *global warming* in e-mails, talks, or other official communications.

In 2014, the city of South Miami came up with an unexpected solution to the lack of concern about sea-level rise. The city commissioners proposed that Florida be cut in half to form two states. The resolution noted that North Florida, where the capital, Tallahassee, lies, is typically above 120 feet in elevation and that South Florida's elevation is 5 feet. Consequently, Tallahassee downplays and underfunds the problem of sea-level rise.

Chuck Watson, a disaster analyst, is quoted in the June 2013 issue of *Rolling Stone* as remarking that "given how much Florida has to lose from climate change, the abdication of leadership by state and federal politicians is almost suicidal—when it isn't downright comical." To date, the *Rolling Stone* article ("Goodbye, Miami," by Jeff Goodell) is probably the most forthright and complete media effort on the probable demise of Miami. An article titled "Miami, the Great World City Is Drowning While the Powers That Be Look Away," by Robin McKie, published in 2014 in the *Guardian,* summarizes the politics of a doomed city led by unfathomably irresponsible politicians. In the February 2015 issue of *National Geographic*, a spectacularly illustrated article titled "Treading Water," by Laura Parker, describes the uncertain future of the city and the lack of significant action by city officials.

According to an article by Robert Meyer in *Business Week*, "Miami has been undergoing a nearly unprecedented surge in real estate construction." This boom goes on despite compelling evidence that the city's very existence is on the brink. A *Miami Herald* article noted that in 2013 there were nine skyscraper multiuse projects under way in greater Miami, most of which were on the waterfront, and four of which were more than 40 stories tall. In addition, in 2013, 20 residential projects were gearing up, ranging from 57 floors to five floors in height. Even MiaSci, the new science museum, is a "waterfront museum" along Biscayne Bay. The two largest developments are the Brickell City Centre and the Miami Worldcenter. The Brickell City Centre is a $1 billion development consisting of four skyscrapers and three other large buildings currently being constructed along the Miami River. When finished, the center will contain apartments, condominiums, a hotel, office complexes, and high-end retail space. About 1,000 condos plus shops and a 1,800-room hotel will make up the $2 billion Miami Worldcenter.

Danielle Paquette wrote in a *Washington Post* article that Miami Beach is encouraging developers to build larger and more costly condominiums. Ironically, Florida has no income tax, and Miami Beach hopes to defend itself from the rising sea with taxes and fees from the condos. Putting more property at risk to raise money to defend it from the rising sea doesn't make sense for Miami or, for that matter, for any other place. In the long term, this approach has scant chance of succeeding. But a variety of adaptation strategies do exist, at least for the short term.

Hal Wanless, Miami's Voice in the Wilderness

No story of Miami and its future in a time of rising sea level would be complete without taking note of the contribution of Dr. Hal Wanless, chair of the geology department at the University of Miami. Wanless has been crying "The sea is rising" for two decades and is largely responsible for the increasing (but largely ignored) recognition of the extreme hazard that the rising sea holds for Dade County. In this effort, he has spoken before a number of political entities, public hearings, schools, and chambers of commerce.

Wanless has stated numerous times, most recently in Goodell's *Rolling Stone* article, that "Miami as we know it is doomed." He argues that instead of spending money on development and port upgrades, we should be buying people out and moving them to higher ground.

One of Wanless's more unusual accomplishments is the establishment of an educational program entitled Empowering Capable Climate Communicators. It is a free class designed for anyone who walks through the door, and it meets on four Saturdays in the spring. Wanless's students range from the occasional high school student to retired professors and even a local suburban mayor. Six or more of his academic colleagues join in the course presentations and discussions, for which no one is paid. The lessons include training the students how to make presentations at hearings and even how to dress for the occasion. The program is a superb example of academia reaching out to the real world as well as an example of the kind of leadership needed in Florida.

The Local Response

Recently Miami-Dade, Broward, Palm Beach, and Monroe counties (all on the southeastern corner of Florida) got together and formed the Southeast Florida Regional Climate-Change Compact. Its primary goal is planning and preparing for the rise in sea level, including establishing adaptation-action areas and finding funding sources. But there is no indication that the compact (or any other political entity in Florida) has realistically faced up to the high-rise problem. That is, what should be done about the barrier islands chockablock with high-rises? Instead, the compact looks to the short term but downplays the dreadfully painful long-term Florida problem (beyond a few decades) of how to respond to dense development on low-elevation land next to a rising sea.

The goals of the Southeast Florida Regional Partnership are similar to those of the Climate-Change Compact but encompass seven counties and include mostly local politicians and officials. The partnership's work also is sometimes controversial, especially when it steps into the climate-change arena. To wit, an Indian River County commissioner who withdrew from the organization declared that the planning blueprint published by the group was a "socialist document."

A letter to the Florida state attorney general, dated November 2013 and signed by 14 groups, including five local tea-party organizations, urged the state to launch an investigation into the alleged official misconduct of elected and appointed Florida officials. The alleged misconduct was the use of what they deemed unproved and perhaps fraudulent science indicating that humans are responsible for the global climate change and that this has resulted in a highly exaggerated threat of global sea-level rise.

Putting this all together provides a very pessimistic view of Miami's future in the era of melting ice sheets and glaciers, expanding ocean waters, and rising seas. The following list summarizes the elements that will have an impact on the doomed city:

- Inundation from sea-level rise with no reasonable possibility of preventing flooding because of the permeable limestone underlying the entire city
- Increased intensity of storms expected to become more powerful because of the warming of the ocean's surface waters
- Increased height of storm surges due to both storm intensification and sea-level rise
- Increased failure of the storm gates across the city's rainfall drainage-system canals as sea-level rise leads to flooding

- Salinization of the water supply due to heavy use and sea-level rise
- Hundreds of immovable high-rise buildings making the potential for extreme damage by sea-level rise very costly and very likely
- The lack of political will to even recognize the problem, much less to make the decisions to effectively respond to it

Not So Easy in the Big Easy

New Orleans is a city of 343,800, with a combined population, including the surrounding communities, of around 1.2 million (in 2010). The city is situated on both sides of the Mississippi River on its great delta. About 5,000 years ago, at the end of the last ice age when the sea level rose to near the current level, the delta began to form as the ice sheets melted. The city now lies in a low-lying flat area with an average elevation of about 2 feet *below* sea level. The highest elevations are 20 feet on the natural levee near the river, and the lowest portions are some 6 feet below sea level in eastern New Orleans. Large areas of greater New Orleans (Plaquemines Parish to the south and St. Tammany to the north) are outside the 133-mile-long dike system that surrounds the city, and of course, the residents there are pleading for the government to extend the walls around them.

According to Virginia R. Burkett, David B. Zilkoski, and David A. Hart of the U.S. Geological Survey, "New Orleans is sinking an average of two inches per decade and it is anticipated that it will sink roughly three feet in the next 100 years." The delta is sinking because of sediment compaction and the extraction of oil and water from the subsurface. The sea level is currently expected to rise about 3 feet, so within this century, the level of the sea around this levied city may effectively rise somewhere around 5 to 6 feet! This is twice the expected rise along the Gulf and Atlantic shorelines away from the sinking delta. As Tim Osborn, a NOAA expert on the delta, exclaimed: "The entire area is sinking faster than any coastal landscape its size on the planet."

Hurricane Katrina and the Storm Surge

As the sea level rises, so does the level of the water in storm surges from hurricanes. The National Weather Service defines a *storm surge* as the "abnormal rise of water generated by a storm, over and above the predicted astronomical tide." It is certain that storm surges will increase in New Orleans as sea-level rise and subsidence combine to raise water levels and as hurricanes intensify owing to the warming of the ocean surface waters.

Hurricane Katrina produced a stunning storm surge of more than 30 feet in Waveland and Bay St. Louis, Mississippi, both on the open-ocean shoreline. Along the New Orleans shores of Lake Pontchartrain, the storm surge was as high as 9 feet and, at the New Orleans airport, was around 12 feet. The funneling effect of the 650-foot-wide Mississippi River–Gulf Outlet (MRGO) shipping channel leading from the city to the open Gulf of Mexico apparently raised the Katrina storm surge by 20 percent and the velocity of the wind-driven water by 100 percent. The U.S. Army Corps of Engineers denies the importance of the MRGO channel to the city's destruction, but its potentially devastating role in a hurricane was predicted well before Katrina's arrival. The recently completed $1.1 billion, 1.8-mile-long Lake Borgne Surge Barrier crosses both the intracoastal waterway and the MRGO channel, with the hope of negating future storm surges enhanced by these navigation channels.

Katrina killed 1,577 people in Louisiana. New Orleans flooded as the levee system surrounding most of the city failed catastrophically. Overall, at least 53 different levee breaks occurred, at least seven of which were I-wall (concrete slabs) failures caused when the slabs separated as they bowed inward with pressure from the rising water.

Protection from Future Storms

The Army Corps of Engineers bears the primary responsibility for the failed levee protection of the city. In a remarkable report entitled "What Went Wrong and Why," the American Society of Civil Engineers criticized the corps for not subjecting the design to the "rigorous external review by senior experts" that is especially important for "life safety structures." To our knowledge, the corps has never followed the practice of using outside project reviewers, and the disaster in New Orleans has not altered its modus operandi.

The breakdown of the levees has been characterized as *the greatest civil engineering failure in the history of the United States.*

The breakdown of the levees has been characterized as *the greatest civil engineering failure in the history of the United States*. The American Society of Civil Engineers noted that the Hurricane Protection System (HPS) (which includes everything intended to protect the city, such as pumps, levees, and gates) suffered from a number of important shortcomings, including the following:

- The HPS's overall design was for a storm smaller than one the National Weather Service would expect as a major storm.
- Because the levees were not armored, they eroded or, in some cases, collapsed when the water spilled over them.
- The crests of the levees were at different elevations (the HPS was not an interconnected system because it was built piecemeal at different times).
- All the pumps intended to keep the city dry failed for a variety of reasons.
- Because several agencies were responsible for the HPS, no single agency could take complete charge.
- Maintenance of the HPS was improper or nonexistent.

The HPS components were intended to safeguard the city, but Mother Nature offers other protection. The hundreds of square miles of wetlands on the delta, many of them located between the city and the sea, offer a poorly understood dampening effect on the storm surge of incoming hurricanes. Unfortunately, Louisiana's wetlands are disappearing at a rapid rate, taking with them their protection for New Orleans. In fact, in the past 80 years, the area of the delta has been reduced by around 1,900 square miles. The wetlands have disappeared for a number of reasons:

- Dams on the Missouri River, which joins the Mississippi, have reduced the essential supply of fresh sediment to maintain the marshes and counteract subsidence.
- The construction of levees along the lowermost Mississippi River has blocked sediment and nutrients from flooding the marshes. Instead, the river and its precious load of sediment and nutrients flow out to the Gulf of Mexico, thereby contributing to the Gulf of Mexico's dead zone off the delta.
- The natural subsidence due to compacting muds and the man-made subsidence due to oil and water extraction from the delta are significant factors. This second cause is, of course, controversial because the petroleum industry downplays the role of oil extraction in sinking the delta.
- The petroleum industry has built an extensive, and marsh-destroying, system of canals throughout the salt marshes to provide access to its production facilities.

Garrett Graves, head of the Louisiana Coastal Protection and Restoration Authority, is optimistic about saving the Louisiana coast, but NOAA officials argue that the rate of sea-level rise in the last half of this century will be so rapid that it will defeat most of the state's costly efforts (using mostly federal funds) to build up barrier islands by dredging sand and to create more marshland by diverting Mississippi sediment–bearing waters to the sinking marshes. Some cost estimates for this approach run as high as $50 billion. Although the situations are not entirely comparable, the barrier islands in front of Waveland and Bay St. Louis, Mississippi, did little to stop Hurricane Katrina's 28-foot storm surges in these communities.

In September 2015, Bob Marshall, writing in *The Lens*, a New Orleans public-interest news site, summarized the unanimous opinions of three winning groups in an international Louisiana Coast design competition. The competing consultants were asked to consider the future out to 100 years. All of them were in agreement that the status quo could not be defended and that retreat in various forms must take place. They concurred that "the required changes are so vast, expensive, and socially disruptive they must take place over several generations." The wetlands and the entire delta will shrink considerably, and outlying communities must eventually be moved. The state should begin planning for this move now. The fishing industry will survive but will experience "dramatic relocations."

We can safely say that generally, local and state government officials are consistently more optimistic than federal officials about the possible success of all the planned engineering risk-reduction activities. One state official noted that if the marsh on the delta could be completely restored, there would be no need for New Orleans to have levees, indicating a wildly unjustified optimism about the dampening effect of marshes on hurricanes' intensity. A complete restoration of the marshes is impossible, but even if they could be restored, the city would still have to contend with storm surges. Moreover, restoring the marshes and building barrier islands (major components of the proposed protection plan) would be very expensive and only temporary solutions, so they would have to be repeated from time to time, just as beaches must be re-replenished.

Taking a cue from the Miami experience, the Army Corps of Engineers is building gates at the end of three drainage canals in New Orleans. Each canal will have diesel-burning pump stations at the gates, as there presumably will be no electrical power during a storm. The gates were part of the $14 billion, once-in-a-100-year-storm-level risk-reduction system that also included repairing and reinforcing the 350-mile system of linked levees, 73 nonfederal pumping stations, and four gated outlets. By far the largest gate is the aforementioned barrier across the MRGO channel.

Risk-reduction projects in the United States generally assume the occurrence of a 100-year storm (a large storm with a 1 percent chance of occurring in any given year), but many scientists believe that in the future, these larger storms will likely come more frequently. Since 1852, 17 Category 3 or higher hurricanes have passed within 100 miles of New Orleans, nine of which were Category 4 or higher, meaning they had winds of 130 to 156 miles per hour. In an article in

Nature Climate Change, researchers from MIT and Princeton University suggested that by the end of the twenty-first century, what are today's 100-year-storms could occur every 3 to 20 years. They also predicted that with climate change, today's "500-year storms" (i.e., a large storm with a 0.2 percent chance of occurring in any given year) could occur every 25 to 240 years.

Risk-reduction projects in the United States generally assume the occurrence of a 100-year storm (a large storm with a 1 percent chance of occurring in any given year), but many scientists believe that in the future, these larger storms will likely come more frequently.

Possibly the greatest threat to New Orleans has nothing to do with storms. It's money. Who will pay for the inevitable and continued required upgrading of levees and pump systems to keep the surrounding waters at bay as the sea level rises? The need to raise levees for a city that already is largely below sea level will come at the same time and for the same reasons as the need to preserve all of America's coastal cities. What will be the national priorities in deciding who will and will not get federal aid? Will taxpayers continue to rescue New Orleans, a very vulnerable city that will be flooded multiple times in the coming decades? Keep in mind that Hurricane Katrina cost $148.8 billion in 2013 dollars. Are New York/Newark, Philadelphia, Boston, and Houston more important and of higher priority than New Orleans? Why should those who choose to live in such a hazardous area as New Orleans continue to be bailed out? Shouldn't they pay for their own storm recovery? These are difficult questions.

The Dutch Dialogues

Senator Mary Landrieu of Louisiana led a delegation to the Netherlands shortly after Hurricane Katrina's landfall to learn from the nation that, by some measures, has the world's most creative water management approaches. The delegation was exploring the idea of changing tactics in New Orleans, by living with the waters surrounding the city and not simply building the levees higher and higher to keep the water out. According to a report from the Institute for Sustainable Communities,

> Although water defines the city in many ways, attempts to manage it have treated this resource as an adversary. Extensive measures to keep water out—complex pumping systems and unsightly levees, flood walls and drainage canals—have ultimately only made the city more vulnerable. The ideas being explored include a new super levee along with neighborhood scale systems like canals, urban wetlands and green spaces for storm water storage and removal of concrete walls that now line drainage canals to provide waterfront access to neighborhood residents.

But this doesn't solve the subsidence problem.

The Dutch Dialogues, as the discussions are called, seem unlikely to result in a new approach. For one thing, the Dutch approach is tailored to a different country, with different priorities, very different resources, in a different natural setting, and with no retreat space if the situation deteriorates. National survival for the Dutch depends on their success in holding back the sea, and their designs are for one-in-500-year events or longer. The United States' national survival is not contingent on the preservation of New Orleans.

National survival for the Dutch depends on their success in holding back the sea, and their designs are for one-in-500-year events or longer.

The Future of a Great City

In sum, New Orleans faces the following:

- A certain future of repetitive and intensifying hurricane activity
- A sea-level rise combined with subsidence that will reduce the levees' effectiveness and increase the height of storm surges, thereby increasing the risk of the levees' catastrophic failure
- The costs of repeatedly recovering from hurricanes, building ever-higher levees, diverting sediment, and carrying out other

risk-reduction policies that will likely be economically impossible, especially because all other American coastal cities at the same time will be responding to sea-level rise and competing for dwindling federal funding

Certainly city and state officials are aware of the problems the city faces, in contrast with those in Miami, whose awareness seems to be either skin deep or totally missing. Several panels and commissions have been set up, and over the years since Hurricane Katrina, science and engineering panels have offered advice and opinions about what went wrong and what should be done in the future. Those panels include those organized by the American Society of Civil Engineers, the Army Corps of Engineers, the National Science Foundation, and the National Academies of Engineering, Science, and Medicine.

Few would argue that the city won't soon be flooded again by some future storm surge. So the solution from the standpoint of individual homeowners is to raise the existing buildings and fill in vacant lots (where homes were destroyed) with raised buildings. Lots of reasonable advice is available to city officials, but the city officials' post-Katrina actions do not provide confidence for the future.

A Tale of Two Cities

The two cities of Miami and New Orleans have quite different hazards, but the principal perils (marked with an asterisk) facing the cities are the same.

NEW ORLEANS	MIAMI
*Sea-level rise	*Sea-level rise
*Low-lying land	*Low-lying land
Hurricanes and storms	Hurricanes and storms
Cost of response	Cost of response
Disappearing marsh protection	*Underlying porous rock
Inadequate levees	Rising groundwater
Land subsidence	Saltwater intrusion

Short of the city's replacing cars with boats, Miami will probably be doomed when the sea has risen 2 more feet. There is no nearby high ground to move buildings to; seawalls and the like won't work; and city and state leaders haven't even agreed that there is a problem, much less started planning how to respond to the coming flood. Making a response all the more difficult, Miami's demise will be slow and gradual: death from a thousand cuts, as the saying goes.

The demise of New Orleans, however, will be either one big catastrophe or a series of catastrophes—that is, big storms. The city awaits the next big hurricane. The potential for extreme damage from storms increases every year as the sea level rises. Unlike the residents of Miami, everyone recognizes the hazards facing New Orleans, and there has been lots of planning, mostly on how to escape the city before the next storm. Long-term planning on the eventual retreat is sketchy at best.

Although much more can be done in the short term to protect these two cities, the survival of both Miami and New Orleans beyond the twenty-first century is in serious doubt.

4

New and Old Amsterdam

New York City and the Netherlands

Because the Dutch were the first Europeans to settle in what is now New York City, the city and the country have had many historical connections. The city was originally called Nieuw Amsterdam and was located on Manhattan Island. New York and the Netherlands also are similar in that New York is the U.S. city best prepared for climate change and sea-level rise, and the Netherlands is the world's best-prepared nation for dealing with those complex problems.

New Amsterdam: New York in the Time of Climate Change

Hurricane Sandy roared into New York City on October 29, 2012, after wreaking havoc from Jamaica up the east coast of the United States. By the time the storm was over, it was the second most expensive hurricane in U.S. history, behind Katrina, with costs estimated at $68 billion. The Red Cross tallied 117 storm-related deaths, with most occurring in New York (53) and New Jersey (34). The most common cause of death was drowning, followed by trauma from being crushed, cut, or struck, and by carbon monoxide poisoning from devices such as generators or grills used after the loss of power.

It might surprise some readers that a major city and world financial center that suffered widespread damage from hurricane-related flooding is nonetheless the United States' best-prepared city for climate change and sea-level rise. After all, Sandy's winds and 13-foot storm surge caused power outages for a million customers, flooded the financial district, and flooded seven tunnels in New York's century-old subway. Entire coastal neighborhoods were devastated. Nevertheless, New York's poststorm actions show that at this point, the city is best situated to respond to the challenges it faces from climate change and sea-level rise (while also revealing that we are in only the earliest stages of the response). One major reason is that New York City's politicians have accepted climate change as a reality and already have started preparing their city for the rise in sea level, whereas many other U.S. cities, like Miami and its politicians, are still either rejecting it outright or just coming to grips with accepting the existence of climate change.

New York's poststorm actions show that at this point, the city is best situated to respond to the challenges it faces from climate change and sea-level rise (while also revealing that we are in only the earliest stages of the response).

Mayor Michael Bloomberg and city officials showed impressive leadership in the months following the storm. Actually, New York already was a leader in addressing climate-change adaptation, having set up PlaNYC in 2008 to address the strategy issue. Part of that strategy was the establishment of the New York City Panel on Climate Change (NPCC), a group of leading climate and social scientists and engineers brought together to develop local climate projections for the city. When Sandy came ashore, New York City already had the experts in place to examine the damage and learn from it. The result, published less than six months after Hurricane Sandy, is PlaNYC's "A Stronger, More Resilient New York," a massive report laying out 250 detailed initiatives, with a $19 billion price tag, that New York can use to address the impending threat of sea-level rise. In addition, the city produced two smaller reports, "Designing for Flood Risk" and "Urban Waterfront Adaptation Strategies." With these three reports, New York has established itself as the leader in

U.S. urban sea-level rise planning, and it has provided an example for other cities to follow.

New York's forward thinking gives the city a head start. In the coming decades, all American coastal cities will finally recognize the need to respond to sea level, and then they will begin to compete for the limited funds to respond to it. It is a certainty that all seaside cities will be seeking federal funding to protect or move their buildings and infrastructure. What's more, the rising seas will affect all coastal cities simultaneously, albeit to differing degrees, and they all, at the same time, will seek money to deal with it. New York already is at the front of the line.

"A Stronger, More Resilient New York" examines in detail many facets of New York's built environment and looks at what worked and what didn't during Hurricane Sandy. It also uses local sea-level rise projections to predict future challenges in the areas of coastal protection, buildings, utilities, liquid fuels, telecommunications, health care, transportation, water and wastewater, and parks. Specific plans of action in all these areas are cited for particular New York neighborhoods and sites. For shoreline protection, the plan calls for a variety of approaches, including replenishing beaches, building dunes, and armoring shorelines by constructing floodwalls and levees, groins, bulkheads, seawalls, offshore breakwaters, deployable floodwalls, and tide gates (sealable pipes like those in Miami that allow storm-surge waters to drain out but that block backflow).

For the time being at least, New York has rejected the idea of building a massive harborwide storm-surge barrier connecting Sandy Hook, New Jersey, with the Rockaway peninsula (known as the Rockaways) on Long Island, New York. The surge barriers would remain open and allow ship traffic and water flow in normal conditions but would close to protect the city during storms. The problem is that the water that is blocked as the storm surge rolls in toward the harbor must go somewhere, and unless long and substantial levees were placed along Sandy Hook and the Rockaways, the floodwaters would do an end run around the barriers and flood the harbor anyway. The storm-surge barrier would also place at an increased risk of flooding those communities on the ends of the adjacent barrier islands, such as Sea Bright and Monmouth Beach in New Jersey and Atlantic Beach and Long Beach in New York. The cities of New Bedford, Massachusetts, and Providence, Rhode Island, both boast floodgates coupled with dikes to prevent flooding, but these relatively small projects would pale in comparison with an effort to construct a harborwide floodgate in front of New York City.

In addition, to rely on such massive infrastructure would also put the city at increased risk of catastrophic flooding when the "big one" comes (similar to the risks that New Orleans faces when the levees fail). That is, rather than reducing development near the shore, massive flood barriers would likely encourage it, so if and when there was a failure, the flooding would be even more catastrophic. Even though New York City is currently rejecting a massive storm-surge barrier, its plan does call for the construction of smaller-scale flood barriers at Newtown Creek and, down the road, at the Rockaway Inlet and Gowanus Canal in Brooklyn. Newtown Creek, a 3.8-mile-long estuary that borders Brooklyn and Queens and empties into the East River, is a U.S. Environmental Protection Agency (EPA)–designated Superfund site—that is, a hazardous-waste site—which should provide an extra incentive to prevent flooding in that area as well as an extra incentive to clean up the site.

At the start of "A Stronger, More Resilient New York," the authors define *resilient* as follows: "resilient (ri-zil-yuhnt), adj. 1. Able to bounce back after change or adversity. 2. Capable of preparing for, responding to, and recovering from difficult conditions. Syn.: TOUGH See also: New York City." This humorous preface to the report is reminiscent of the boastful responses that politicians typically trot out following disasters: We will bounce back—stronger, better, tougher. It is good to be tough, but even better to be smart. New York's plan is smart but could be smarter. The introduction to the report states that the "city cannot, and will not, retreat." That is a tough but ultimately absurd stance. Bloomberg also publicly

> Ultimately, if one were not to retreat at all, the city would have to build 500 miles of dikes, levees, and massive seawalls.

proclaimed at the Duggal Greenhouse in Brooklyn on June 11, 2013, when the report was released, that we "cannot and will not abandon our waterfront" and that we "must protect it, not retreat from it." A strong statement indeed, and just what you would expect from a politician. But it is not smart or realistic. Ultimately, if one were not to retreat at all, the city would have to build 500 miles of dikes, levees, and massive seawalls.

As well prepared as it already is, New York City, like every other seaport city, eventually will have no choice but to retreat to some extent. The sooner that seaside cities accept this, the better prepared they can actually be for sea-level rise. Unfortunately, we continue to be impeded by recalcitrant climate-change deniers. Maybe it is too early for politicians to acknowledge that we will indeed have to move back from the sea or otherwise it will envelop us.

It is this state of political (non)reality that brings us to a major weakness of New York's planned response to sea-level rise. While the cleanup of Sandy was still under way, Bloomberg convened the Special Initiative for Rebuilding and Resiliency to study the impact of the storm and assess the risks from climate change. This committee is tasked with analyzing this threat in the medium term (2020s) and the long term (2050s). New York City is basing its climate adaptation on "long-term" predictions set for midcentury—just a few decades away. This is woefully inadequate because the seas will not stop rising by 2050. In fact, the rise in sea level is expected to speed up swiftly in the near future; currently, it has really just begun. New York has many structures that are hundreds of years old, and many of the buildings and other structures that will be built under the city's new plan will probably exist for decades and centuries beyond 2050. Why not create new infrastructure and buildings following codes that are designed to adequately protect them against the kinds of conditions they most likely will face during their expected lifetimes

> The United States is used to thinking in terms of "100-year floods." Instead, why not emulate the Dutch, who, when feasible, design their structures to withstand 10,000-year storms?

(and not just the next few decades)? Nationally, the United States is used to thinking in terms of "100-year floods." Instead, why not emulate the Dutch, who, when feasible, design their structures to withstand 10,000-year storms?

In contrast to New York's plan, the city of Miami, which is in much more immediate danger from sea-level rise than New York, has not moved much beyond talking about sea-level rise. Miami continues to build skyscrapers at soon-to-be-flooded low-elevation sites. As we pointed out in chapter 3, Miami's $1.1 billion Brickell City Centre and the $2 billion Worldcenter developments will add new skyscrapers to the city's horizon. In North Carolina, the legislature initially stated that sea-level rise would not be allowed to accelerate (thereby making it "illegal"), but after much ridicule, it has now retreated from that position. A science panel in North Carolina is again looking at sea-level projections, although by law the panel can make projections no further than 30 years into the future. Shortsighted indeed!

New York may be ahead of most American cities in the planning process, but the city's approach still has problems. Is the choice to limit planning for a major city, for what can be considered only the short term, one that was made for political expediency? Is it unpalatable to look far enough into the future and to plan for the much higher sea level that will be upon us down the road? In 2015, the New York City Panel on Climate Change has extended its projections through the year 2100. This extension is an improvement but will be meaningful only if the plan is altered to reflect the higher seas and the greater risks that will surround the city by that time. We should view the end of the century as the short term, however, in regard to climate change and sea-level rise. A study published in the July 2013 *Proceedings of the National Academy of Sciences* asserts that for every 1-degree Fahrenheit rise in temperature caused by global warming, we lock ourselves into a 4-foot rise in global sea level. When New York finally abandons its myopic approach to sea-level rise and looks beyond just a few decades or even the end of this century, the city's current refusal to consider retreating from the waterfront will look even more foolish.

The topography of New York offers it some advantages in regard to retreat that many other cities lack. While the city does have many miles of coastline/shoreline occupying glacial outwash plains, it also features hills and high bluffs, some of which are anchored on bedrock. In other words, New York has high ground to which it can retreat. New York could relocate people to high-density residences out of danger from flooding. When it comes to topography, low-lying Boston and Miami and the Netherlands are not as blessed.

New York City is considering a number of projects to protect Lower Manhattan from further flooding, including Seaport City, a massive multipurpose levee that would simultaneously raise the shoreline to 19 feet above current sea level while expanding Manhattan some 500 feet into the East River. This project was deemed the most financially feasible of the proposed projects to protect Lower Manhattan because it would allow residential and economic expansion to offset the construction costs, which could be billions of dollars.

New York City's saving grace may be its plan to regularly revisit its climate-change predictions. New York City law requires that the New York City Panel on Climate Change meet twice a year, advise the city and the Climate-Change Adaptation Task Force on scientific development, and update its climate projections at least every three years. We can only hope that as the depth of the problem of sea-level rise becomes clearer, New York will modify its plans and allow for managed coastal retreat. At least two of the scientists who worked on the NPCC report recognize that retreat is in the city's future. As Philip Orton, a physical oceanographer, has stated, "At some point, I suspect we'll have to abandon at least some areas," and Klaus Jacob, a geophysicist at Columbia University, agrees with him. According to Jason Diaz, Jacob favors a "managed retreat," converting buildings in flood zones into green spaces and concentrating buildings on higher ground. Jacob is pushing New York in the right direction, and the sooner the city accepts this inevitable solution, the sooner it can ensure the safety of its citizens and avoid the massive costs associated with storm recovery.

Jacob has proposed some visionary ideas for New York's future, including sealing off the basements and first floors of buildings in flood zones and using the lower levels only for parking in order to minimize damage from flooding. He also suggests building an elevated walkway, like the city's celebrated High Line, connecting buildings and allowing for safe transportation during and after flood events. More significant, Jacob calls for converting flood-prone areas into green spaces and relocating people and businesses to higher ground. This would allow the city to accommodate floodwaters with minimal damage. Of course, this is a radical and costly idea. In his June 2013 speech at Duggal Greenhouse, Bloomberg noted that more than 500 million square feet of New York City buildings fall within the Federal Emergency Management Agency (FEMA)'s current 100-year flood maps, including the homes of nearly 400,000 people, plus more than 270,000 jobs. As much as one would like to discount Jacob's vision as dystopian science fiction, the sad fact is that along with our inability to muster the will needed to reduce greenhouse-gas emissions, we can expect continued warming in the atmosphere, continued and increased melting of the world's ice sheets, and continued rising of the oceans. Jacob's proposal of abandoning shorelines and elevating emergency paths may prove to be prophetic.

New York City did not start to move back from the shore until Governor Andrew Cuomo stepped in and offered a buyout of hazardous coastal lands. Using federal funds authorized for recovery from Hurricane Sandy, Cuomo, through the state agency NY Rising Community Reconstruction (NYRCR) Program, is targeting properties vulnerable to flooding. The program creates bonuses to encourage participation. Specifically, New York's buyout plan offers a 5 percent bonus over the pre-Sandy market value of properties in certain vulnerable neighborhoods if the residents relocate in the same county. The state also is offering an additional 10 percent bonus to certain high-risk areas to encourage greater participation in the buyout in order to avoid future disasters.

The 23 Staten Island residents who perished during Hurricane Sandy account for nearly half of all the Sandy-related deaths in the state of New York. Fittingly, the neighborhood buyout program has focused on three Staten Island neighborhoods: Oakwood Beach, Ocean Breeze, and Graham Beach. The buyouts have proved popular in those neighborhoods that were hit in quick succession by a nor'easter in 2010 and then Hurricane Irene (2011), followed by

Hurricane Sandy (2012). The state has purchased or offered to purchase approximately 130 homes in Ocean Breeze, 205 homes in Oakwood Beach, and, at the time of this writing, had announced plans to purchase 200 properties from the Graham Beach area. NYRCR set aside $129 million to buy the 205 properties in Oakwood Beach and an expected 114 more parcels to be bought in late 2014. Meanwhile, the Army Corps of Engineers plans to turn Oakwood Beach into a "belt retention pool" of wetlands to help absorb future storm surges, at a cost of $160 million.

NYRCR is offering "buyouts" with the aforementioned offers of up to 15 percent above the pre-storm value to buy and demolish homes in order to create coastal buffer zones. The program also is offering "acquisitions," in which homes outside the designated buyout areas are purchased at the pre-storm value with the intention of replacing them with more-storm-resilient buildings. These would be older houses built before the advent of modern building codes. On nearby Long Island, NYRCR has identified an area of Suffolk County as a buyout area where rebuilding would not be permitted after the houses were razed. NYRCR has offered to buy 470 homes in Nassau and Suffolk Counties and is reviewing applications from another 469 homeowners. In total, New York is dedicating $600 million in federal disaster relief funds for buyouts. Next door in New Jersey, the state plans to use $300 million in federal disaster recovery funds to purchase around 1,000 properties in areas affected by Sandy, a drop in the bucket (or ocean) perhaps, but it is a start.

Old Amsterdam: The Dutch and Sea-Level Rise

Any modern discussion of the Netherlands' approach to battling the threat of ocean flooding must begin with the Watersnoodramp, the 1953 North Sea Flood. Of course, the Dutch have been constructing dikes, mounds, and other flood-control barriers for centuries. The iconic Dutch windmills were, in fact, pumps to keep water out of the *polders*, the low-lying land between dikes subject to subsidence (sinking). Interestingly, Dutch water boards, the committees responsible for water management, were democratically elected as early as the

> The iconic Dutch windmills were, in fact, pumps to keep water out of the *polders*, the low-lying land between dikes subject to subsidence (sinking).

thirteenth century. The storm in 1953 coincided with a spring tide, resulting in an extremely high combination of tide and storm surge estimated to reach as high as 18.5 feet. Flooding and deaths also spread across the United Kingdom and Belgium, but the Netherlands was especially devastated by the numerous breaches in the dikes, resulting in more than 1,800 fatalities, as well as an estimated 30,000 animals drowned and tens of thousands of buildings damaged.

This unprecedented flood disaster prompted the creation of the Delta Commission, a committee whose work culminated in the Delta Works, a nationwide flood defense designed to combat both storm surges and flooding from rivers. The Netherlands lies on the combined deltas of three major rivers, the Scheldt, Meuse, and Rhine. This expanse of waters and wetlands, all at low elevation, makes the country particularly vulnerable to flooding and the effects of sea-level rise. Add to this the fact that approximately one-quarter of the nation lies below the current sea level and about half the nation is threatened by flooding rivers and storm surges, and you can begin to understand just how great a problem sea-level rise is for the Dutch. The Netherlands is unlike the United States, where most of the states and land are not adjacent to the coast and the residents of landlocked states may balk at spending tax funds to defend the coast. Instead, to the Dutch, the threat from climate change and sea-level rise is so extreme and so real that they are united in their devotion to tackling these problems.

> To the Dutch, the threat from climate change and sea-level rise is so extreme and so real that they are united in their devotion to tackling these problems.

Although the Dutch, renowned for expanding and hardening their shoreline, are not about to abandon their ambitious engineering feats in the time of rising seas, there has been a significant shift in their relationship to water. As the scientific evidence of climate change mounted, the Dutch formed a second Delta Commission in 2007. The commission's findings were detailed in the 2008 report "Working Together with Water," showing that it recognized that the Netherlands' future lay in a strategy of both flood protection and sustainability. The Delta Commission established its Delta Programme to guide the nation in adapting to climate change. The commission anticipates a sea-level rise between 6 and 12 feet by 2100 (taking subsidence into consideration). It also expects greater river flow during the winter due to milder winters and less discharge during summer due to drier, warmer temperatures. A major concern is the threat of saltwater intruding into the freshwater supply. In order to address these expected troubles, the Dutch anticipate spending a whopping €1.2 billion to €1.6 billion *each year* through 2050 and slightly less than that through 2100.

What will the Dutch do with all that cash? Well, for starters, they will increase the flood protection of the diked areas by a factor of 10, in some areas by constructing Delta Dikes—that is, dikes so wide and massive that they are unlikely to fail. The Delta Programme also calls for a cost-benefit analysis for new urban development requiring that the costs of local decisions not be passed on to society as a whole. Rather, these costs must be borne by those who benefit from the plans themselves. This disincentive will certainly discourage construction in risky areas. In addition—and this is a prime example of the newer, "softer" approach—construction outside the dikes must not impede the rivers' discharge capacity or the lakes' future water levels. Measures are to be taken to accommodate the expected increase in the discharge of river waters, including the purchase of land to allow the storage of excess water during times of high discharge. In other words, some of the floodplains will be allowed to return to their natural function of dealing with excessive water. Instead of being destructively inundated by rain and river waters during heavy storms, the Dutch will retain some of the waters temporarily and then release them when the storm passes. Because saltwater flooding is so much more destructive, when the storm barriers are up to prevent inundation from the sea, the Dutch can temporarily store river waters. The Dutch are employing clever ways to temporarily store freshwater, including parks designed to double as water retention ponds and, in Rotterdam, a parking garage with a large built-in subterranean rainwater storage tank.

As we noted earlier, the Dutch are also mindful of the threat of saltwater intrusion into their water supplies, so they are planning to expand their freshwater reservoirs wherever possible. But they also are anticipating the need to eventually restore the tidal dynamics at the storm-surge barrier in the Eastern Scheldt, recognizing that sea-level rise will eventually defeat the barrier. (The barrier currently can withstand a 3-foot rise in sea level.) The Dutch realize that the loss of the intertidal zone caused by the construction of the sea barrier will significantly reduce the accretion of sand and thus erode the protective tidal flats. The plan is to replenish the sand in the immediate future while increasing the flood protection in the lands adjacent to the coast in anticipation of the surge barriers becoming obsolescent. Thus, even in the presence of massive infrastructure designed to hold back the seas, the Dutch are planning some form of limited retreat.

The Dutch hope to take advantage of massive offshore sand deposits to help maintain their shorelines in their current location for as long as possible despite sea-level rise. Indeed, sand replenishment as flood control is a major theme emerging from the Dutch plans. The Dutch are blessed with both massive supplies of sand and a massive supply of money. They are testing a new approach called the Sand Engine, which does not use the traditional methods of depositing sand directly on the beach or directly offshore. Instead, the Sand Engine entails building, using offshore sand, a half-mile seaward hook-shaped peninsula in front of the beach. The idea is that the body of sand protruding seaward will erode and provide a continuous supply of sand to the adjacent beaches. The back side of the hook will create a small lagoon sheltered from the waves to serve as a habitat for fish and a recreational spot for beach visitors. Regardless of whether this engineering innovation is effective, the Dutch will be increasing their replenishment of beach sand as a cornerstone of their flood-protection plans.

One positive effect of sea-level rise and of the Netherlands' long history combating flooding is that the Dutch are in great demand internationally by communities concerned with these matters. This is not a new development. The Dutch advised officials in Galveston, Texas, following the city's devastation from a hurricane in 1900. The Dutch have also helped build storm protections in various locations around the world, including in St. Petersburg, Russia, and they are also behind major plans to build a static defense in Jakarta, Indonesia (see chapter 5).

One thing is certain: the Dutch engineering approach cannot possibly work for America's entire coast. Some large cities could construct storm barriers or large dikes to protect themselves from the rising seas. But the American coastline is huge compared with the Netherlands' coastline, and it is financially infeasible to armor the entire American coast, not to mention the environmental implications. The problem is that the Dutch want to hold the shoreline in place even at great cost. The Netherlands is a small, crowded country situated on the delta of three major rivers with a large percentage of its land below sea level. The length of the Dutch coastline is approximately 1,139 miles. Florida's coastline alone is 1,350 miles long. American barrier islands measure 3,000 miles, and the total U.S. coastline—including the Atlantic, Gulf of Mexico, Pacific, and Arctic coasts—is estimated at 88,633 miles. Another major difference between the United States and the Netherlands is that any retreat in the Netherlands must be relatively minor because the Dutch simply lack the land to which they might retreat. They have no choice but to armor the coast while strategically abandoning some areas to water. In contrast, the United States has plenty of land to which communities can retreat. It therefore makes far more economic sense to retreat from coastal areas in the United States than to fight ever-increasing sea levels to keep communities locked in place. Moreover, the cost of following the Dutch approach would likely cripple the U.S. economy and would certainly fail in the long run.

5
Cities on the Brink

The financial exposure that a rise in sea level brings to coastal cities is obviously an order of magnitude greater than the financial vulnerabilities of low-lying, tourist beachfront developments. Coastal urban areas teem with major trading ports, are crowded with valued infrastructure, and draw in large human populations that sprawl from the center of cities miles and miles into their suburbs. Coastal cities are typically built on sedimentary strata in coastal plains, in river deltas, or on engineered fill. They may be prone to land subsidence from the excessive groundwater withdrawals needed to supply an ever-growing populace. Particularly exposed residents may be crowded into slums.

In urban areas, rising seas threaten critical power generation and distribution facilities because the salt in the seawater corrodes equipment, thus undermining its strength. Floods associated with sea-level rise release chemicals and oil into streets and homes. Floodwaters overwhelm storm and wastewater sewer systems and spill sewage into streets and rivers and onto beaches. When sea-level rise hits the world's cities, its health effects are more likely to fall on the poor, the old, the very young, and people with chronic medical conditions.

Cities and Human Health

Extreme events in crowded urban areas will disrupt health care, public-health services, and the availability of fresh food and water, thereby exacerbating underlying health conditions and the ability to control infectious diseases. The rise in sea level and its associated floods will compromise food security and increase malnutrition, spreading infectious diseases, causing food poisoning, and exposing people to pathogens as sewage systems overflow. These changes will likely contribute to mental stress and post-traumatic stress disorder (PTSD). The impact in urban areas will be greater in those less wealthy countries lacking adequate infrastructure and capacity in health services and public health.

The 2014 World Health Organization's "Report from the Intergovernmental Panel on Climate Change (IPCC)" is the fifth in a series of WHO reports on the effect of climate change on health. The report outlines with "high confidence" the long-term risks to human health related to climate change and sea-level rise. In urban areas, the most likely risks are

- Death, injury, ill health, or disrupted livelihoods in low-lying coastal zones and small-island developing states and other small islands due to storm surges, coastal flooding, and sea-level rise
- Severe ill health and disrupted livelihoods for large urban populations due to inland flooding in some regions
- Systemic risks due to extreme weather events, leading to the breakdown of infrastructure networks and critical services such as electricity, water supply, and health and emergency services
- Mortality and morbidity during periods of extreme heat, particularly for vulnerable urban populations and those working outdoors in urban or rural areas
- Food insecurity and the breakdown of food systems linked to warming, drought, flooding, and precipitation variability and extremes, particularly for poorer populations in urban and rural settings

The WHO reports have been criticized by various deniers, including the Nongovernmental International Panel on Climate Change

(NIPCC). The NIPCC has countered by noting potentially positive outcomes of climate change and sea-level rise, such as fewer cold weather–related deaths. But even this group acknowledges that sea-level rise will make existing poverty-related health problems even worse.

According to the United Nations' predictions in its 2013 "World Population Prospects" report, by 2050 one in six people will be age 65 or older, which is double the proportion today. Between 2010 and 2050, this population in the United States is expected to increase by 111 percent, and by 181 percent in the world (1.5 billion worldwide). About 54 percent of the world's population, a number that is expected to increase to 66 percent by 2050, now live in urban areas. Much of that expected growth will be in developing countries.

In a very thorough study for the Organization for Economic Cooperation and Development (OECD), Robert Nicholls and his colleagues compared the relative exposure of the world's largest port cities to flooding from rising seas and storm surges, both for today and projected into the year 2070. (The OECD is an international organization whose member nations share information on trade and economic growth.)

The report ranks cities' vulnerability to a once-in-a-100-years surge-induced flood based on the economic value of their port and city assets (buildings, transportation, and utility infrastructure). The OECD study doesn't account for flood adaptations that could change the impact of the future events. Using this lens, 60 percent of the world's assets that are exposed to sea-level rise and flooding are found in three wealthy, developed countries: the United States, Japan, and the Netherlands (table 1). The report then projected how the top 10 ranked cities would change by the year 2070. Over the coming decades, the growth and economic development of the largest cities in Asia are key driving factors in how this list would change.

Sixty percent of the world's assets that are exposed to sea-level rise and flooding are found in three wealthy, developed countries: the United States, Japan, and the Netherlands.

TABLE 1
Asset Exposure of World Port Cities

Top 10 Cities, 2005	Top 10 Cities, 2070
Miami, Fla., United States	Miami, Fla., United States
Greater New York City, N.Y., United States	Guangdong, China
New Orleans, La., United States	Greater New York City, N.Y., United States
Osaka–Kobe, Japan	Kolkata (Calcutta), India
Tokyo, Japan	Shanghai, China
Rotterdam–deltaic Rhine River, Netherlands	Mumbai (Bombay), India
Nagoya, Japan	Tianjin, China
Amsterdam, Netherlands	Tokyo, Japan
Tampa–St. Petersburg, Fla., United States	Hong Kong, China
Virginia Beach, Va., United States	Bangkok, Thailand

Source: Data from Susan Hanson, Robert Nicholls, N. Ranger, et al., "A Global Ranking of Port Cities with High Exposure to Climate Extremes," *Climatic Change* 104 (2011): 89–111.

TABLE 2
Population Exposure of World Port Cities

Top 10 Cities, 2005	Top 10 Cities, 2070
Mumbai (Bombay), India	Kolkata (Calcutta), India
Guangzhou (Canton), China	Mumbai (Bombay), India
Shanghai, China	Dhaka (Dacca), Bangladesh
Miami, Fla., United States	Guangzhou (Canton), China
Ho Chi Minh City, Vietnam	Ho Chi Minh City, Vietnam
Kolkata (Calcutta), India	Shanghai, China
New York City, N.Y., United States	Bangkok, Thailand
Osaka–Kobe, Japan	Yangon (Rangoon), Burma
Alexandria, Egypt	Miami, Fla., United States
New Orleans, La., United States	Hai Phong, Vietnam

Source: Data from Susan Hanson, Robert Nicholls, N. Ranger, et al., "A Global Ranking of Port Cities with High Exposure to Climate Extremes." *Climatic Change* 104 (2011): 89–111.

Using a different lens, the report also ranked vulnerable cities by the exposure of populations of 1 million or more (table 2). The study's authors identified 136 vulnerable cities. In today's population estimates, the total human population at risk in these 136 cities is around 38.5 million. By 2070, that vulnerable population is predicted to grow to 150 million. Nineteen of the 136 cities are in Africa, which, for a number of reasons, is particularly ill-equipped to respond to sea-level rise. They lack organized data to document the rise, the economic means to respond, and flood management capabilities.

Most of the top 10 vulnerable cities in regard to population are located in deltaic coastal environments. The list is almost equally split between developed and developing countries. Today, Mumbai has the largest population exposed to increased sea-level rise and coastal flooding, but by 2070, Kolkata will have the most vulnerable population in a major coastal city.

The Great Garuda in Jakarta, Indonesia

Bill Tarrant's "In Jakarta, That Sinking Feeling Is All Too Real," in the 2014/2015 Reuters series Water's Edge: The Crisis of Rising Sea Levels, discusses Jakarta, Indonesia, where sea-level rise already is at a crisis stage. Jakarta is considered a megacity, which, according to the United Nations, is a metropolitan area with a population of 10 million or more people. Jakarta is a melting pot of cultures, a place of great diversity. It is modern with many high-rise buildings and a center of finance, but also with many poor areas where people live in shelters made out of whatever materials they can find.

Throughout history, Jakarta has been ruled by a number of countries. It was called Batavia when it was part of the Dutch colonial empire in the seventeenth and eighteenth centuries and then became Djakarta during the brief occupation by Japan during World War II. Today Jakarta's population is estimated to be 28 million people. The Dutch influence is still apparent today in Indonesia's choice to go for the hard-engineering solutions to calm the rising waters.

Jakarta is located at the mouth of the Ciliwung River on Jakarta Bay, an inlet of the Java Sea. The Ciliwung River is one of a number of heavily polluted rivers flowing through the larger city, and land subsidence has interrupted their ability to flow freely into Java Bay. The river waters are stored temporarily in reservoirs and are pumped or channeled into canals to make the last leg of their journey. The city has just one wastewater-treatment plant that serves the central business district. Most of the population instead uses septic tanks or merely dumps waste into neighborhood sewers that flow into the canal system. Jakarta thus is living a crisis driven by its population, history, sea-level rise, and subsidence. No other city is sinking faster than Jakarta, whose annual subsidence ranges from 3 to 6 inches a year. Even the decades-old seawall built to keep the Java Sea from flooding into the city is sinking.

A monsoon storm in 2007 that coincided with a high tide overwhelmed Jakarta's seafront and pushed a wall of water from the bay well into the city. Nearly half of Jakarta was covered by as much as 13 feet of muddy, filthy water. A half-million people were left homeless, and at least 76 people were killed. Economic damages were estimated at nearly $550 million. Although this wasn't the first or the last flood for the city, it spurred the government to act.

A task force was formed to consider the city's options, including retreat or abandonment of parts of the greater city. Another serious storm and flood occurred as the task force debated, pushing them to take bold action. Jakarta has chosen to gamble its future on walling off inundation from the sea, with an offer of $4 million from the Netherlands for a feasibility study.

Jakarta is trying to ensure its own future for the urban agglomeration by building the "Great Garuda," a protective wall and reservoir system bearing a resemblance from the air to the Hindu bird creature. (The Garuda, a national symbol of Indonesia, is from a Hindu myth about a powerful bird-god. Half-man, half-eagle, the Garuda steals an elixir from the gods that promises immortality.) The cost of the Great Garuda is projected to be $40 billion, and it will take 30 years to build. It is, however, a static, in-place development, which means that in the long run of several decades, Jakarta will again be threatened and either will be "saved" by massive seawalls or parts of the city will be forced to move.

The Great Garuda will be a 15-mile-long wall that will be constructed a few miles out to sea and will form a reservoir behind it to collect floodwaters. It will be equipped with a pumping system and retention areas to defend against the intrusion of seawater. Parts of the city will be relocated on top of a series of artificial islands to be built as part of the project. Other aims of this project are to provide clean drinking water to Jakarta, clean up the river systems, and improve the lives of millions of people.

The Great Garuda is truly a bold idea and a mammoth undertaking. Part of the long-term financing is expected to come from selling upscale real estate from the project. Because this is a country where corruption is the norm, there will be large enough amounts of money in this project to attract greed. Meanwhile, construction of a $2 billion seawall has begun inside the old wall, with the hope of buying time to work through the approval and financing process for the larger project. In the first phase, the plan is to add pumping stations and increase the height of existing dikes.

Urban Lowlands: Cities Built on Landfill

Cities built on landfill during growth and expansion are particularly at risk from a rise in sea level. Fills are typically constructed in lagoons or estuaries rather than the open ocean. Almost without exception, these reclamation projects extending buildable land into the sea or bay end up with very low-elevation land. An exception is the large bodies of sand that can accumulate updrift of long jetties that extend seaward from inlets, built to provide a protected navigation channel to the open ocean.

Sullivan's Island, South Carolina, has a rock revetment seawall that never fails to surprise tourists. It is located in the interior of the island well back from the sea and marks the location of the shoreline before the Charleston jetties were built in the 1890s. The shoreline there has moved substantially seaward, far enough to allow rows of houses to be built over what was once seafloor (and what will one day once again return to seafloor).

Another example of construction on sand accumulated by a jetty is the 31-story Point of Americas condominium at the southern end of Fort Lauderdale, Florida. But the greatest of all developments on sand amassed by engineering structures is north of the Santa Barbara, California, breakwater where a huge amount of sand has been trapped on what used to be seafloor. Today, a sports stadium, parking lots, and various buildings have been constructed there. The price paid for trapping sand for this artificial land was increased erosion for a dozen sand-starved shoreline miles to the south (downdrift).

Widening the lagoon sides of barrier islands is another use of landfill to expand cities. Islands in New Jersey were enlarged in the 1940s and 1950s by bulldozing dune sand from the island to cover the marshes behind the island. This process of widening the islands simultaneously removed the sand dunes and lowered the elevation, both of which once offered protection against storm surges. Today, on a number of New Jersey's barrier islands, the only high points are the frontal dunes made by bulldozing sand from the beach.

Each city has a different story, history, and geologic framework (table 3). Many of the urban fills are used to fill in wetlands, including tidal flats and salt marshes. For example, Manhattan is built on large areas of filled salt marshes, as well as fills extending (typically 1,000 feet) into the Hudson River to the west and the East River to the east. Twenty percent of Singapore (52 square miles) is built on top of fill, and the country is in the process of adding another 20 percent. Unfortunately, Singapore often uses beach sand from nearby countries. Likewise, much of downtown Tunis, Tunisia, is built on fill extending far out into the Lake of Tunis lagoon. This was begun when the Romans started filling in this lagoon after they conquered nearby Carthage. In addition, Monaco is 20 percent larger than it once was; Bahrain is 76 percent larger; and Macao has grown more than 100 percent.

TABLE 3
Cities with Substantial Fill Areas

Boston, Mass., United States	Dhaka (Dacca), Bangladesh
Singapore	San Francisco, Calif., United States
Dutch cities	Hong Kong, China
Helsinki, Finland	Manhattan, N.Y., United States
Rio de Janeiro, Brazil	Wellington, New Zealand
Mumbai (Bombay), India	St. Petersburg, Russia
Tokyo, Japan	Tunis, Tunisia
Newark, N.J., United States	Macao, China
Beirut, Lebanon	Monaco

The largest fill in North America is the 5,250-acre project in Boston, an area that amounts to about one-sixth of the whole city. The city was established in 1630, and major fills were carried out in 1830, 1845, 1865, and 1890. Some of the fill came from the excavation required for construction of a wall. Additional fill was dredged from Massachusetts Bay or shipped in by railroad cars from the suburbs after the workers used up the materials they had obtained from flattening hills inside the city.

The low-elevation fills make the downtown section of Boston particularly susceptible to sea-level rise, and several discussions have been held about the appropriate response. The view of Gina Ford, a landscape architect with Sasaki Associates, is particularly intriguing, as it involves little shoreline engineering and emphasizes working with water. These are her ideas:

1. Collaborate on a regional plan. What one community does can affect the adjacent communities. Many solutions will cross jurisdictional boundaries.
2. Make room for water in the city. This involves an approach much like the new Dutch response, in that water can be accommodated; it should not be fought and kept out. Plans for the city include areas to be periodically flooded with storm water and higher tides. Flooded areas can be public parks, certain roads, and underground cisterns.
3. Establish new building regulations. Future buildings must be designed with flood protection and accommodation in mind. Old buildings should be retrofitted. Once this has been done, making room for water in the city may be easier.
4. Make a different sea edge for the city. Alternatives to the current, hard-stabilized (mostly seawalls) city edge could include floating neighborhoods, floodable open spaces, terraced public spaces, and parks that can absorb water.

Decisions made by governments today may be to wait and see and then to react to sea-level rise. These decisions typically are manifested as insurance incentives to rebuild, permit decisions that encourage development in high-risk areas, and consideration of climate-change deniers' demands for more certainty and perfect data in science. Conversely and preferably, we can adapt, finance, plan now, expect to fail occasionally, and use our experience to improve and gain resilience. Today's responses to sea-level rise in cities will surely include some tried and true hard-engineering fixes (seawalls, dikes, and dams), as well as some other hard solutions (storm-surge barriers), soft-engineering solutions (wetland restoration, beach replenishment), and, the most difficult solution of all, intentional retreat from coastlines. Because elected government officials, other public officials, regulators, businesses, and citizens are not used to the large-scale cooperation that this will require, we need new models for cooperation between governments and concerned citizens.

6

The Taxpayers and the Beach House

Depending on whom you believe, the Outlaw family's house in Nags Head, North Carolina, has been moved back either three or five times, a total of 600 feet over a time span of 120 years. The earliest moves were made by mules dragging the house back on top of logs.

A different approach is used today by barrier-island dwellers along Colombia's Pacific coast. Here, the houses in the small villages often sit very close to rapidly eroding shorelines that retreat in spurts corresponding to the frequent local earthquakes. The inhabitants persist in living here in such hazardous locations in part because the sea breeze helps blow away some of malaria-carrying mosquitoes. Rather than moving whole houses through the rain forest when they are threatened by erosion, they are moved back piece by piece. The houses' walls are made of light wooden panels that two men can carry to a new house site. The same two workers can then move a house out of danger in just a couple of weeks (after the new pilings are put in). Both the Outlaw house and the Colombian approaches to retreating from the shoreline are carried out without any government funding, regulations, or advice. The moves are made at the owners' expense or through their efforts, and in both cases, the beach remains intact, albeit farther inland.

Today, the shorelines of the United States and many other countries are jammed with large, multistory houses that for all practical purposes are immovable and are on lots so small that there is no space to move them back. Now the sea level is rising, and the government is very much a part of the picture. Ironically, however, the actions of the U.S. government have encouraged dangerous and environmentally damaging shoreline development.

Mitigation: A Key Tool for Responding to Sea-Level Rise

The Federal Emergency Management Agency (FEMA) defines *mitigation* as the effort to reduce loss of life and property by lessening the impact of disasters. Accordingly, the government mitigates damage by identifying risks and encouraging state and local communities to take steps to reduce these risks. Managed coastal retreat can be considered the ultimate mitigation program because it removes people and property from harm's way.

Congress established the National Flood Insurance Program (NFIP) in 1968 to provide subsidized coverage in areas prone to flooding. Private insurance was either expensive or unavailable (most homeowner policies exclude coverage for floods), so the idea was to couple subsidized flood insurance with the promotion of mitigation efforts, such as building standards like elevated structures. Therefore, the federal money comes with strings attached. For example, in order to get flood insurance, you may have to place your house on pilings to allow water to pass beneath it. This program was progressive insofar as it introduced the concept of community-based mitigation in flood areas. In exchange for subsidized flood insurance, communities had to pass ordinances restricting further development in flood areas. Initially this was a voluntary program, but participation in the NFIP mushroomed after the purchase of the flood insurance was made mandatory for homeowner loans backed by federal mortgages in flood-prone areas. The NFIP's goals are preserving and restoring floodplains, limiting development in flood-prone areas, and implementing better building standards. Communities that exceed the minimum flood-ordinance requirements are rewarded with lower insurance premiums through the NFIP's Community Rating System.

THE CASE OF WAFFLE HOUSE

What can a business like Waffle House tell us about the severity of a storm? More than 1,500 Waffle House restaurants are scattered throughout the Southeast, where hurricanes are common and the restaurant chain is recognized as a model for its crisis management. Waffle House has a manual of instructions for opening after a disaster and keeps portable generators on hand. In addition, it tracks storms and moves supplies in anticipation of damage. The restaurant chain even has an emergency menu designed for speed and efficiency that does not include waffles (too much electricity) or bacon (too much time and grill space). Waffle House even is equipped to house employees near restaurants during poor weather. FEMA's director, Craig Fugate, coined the term "Waffle House Index" as an informal test to quickly determine the severity of natural disasters. Fugate's Waffle House Index runs from green to yellow to red. It's green if a Waffle House is open as usual with electricity, suggesting low damage. Yellow means a Waffle House is running on a limited menu with electricity from a generator and indicates moderate problems. Red means a Waffle House is closed, suggesting severe damage. During a visit to Gulfport, Mississippi, on the tenth anniversary of Hurricane Katrina, Fugate recalled that after Katrina, storm-response crews sent back photos showing that the "Waffle Houses were not only red; they were only [concrete] slabs," pointing out that even though the national media were focusing on New Orleans, the Mississippi Gulf region also had been devastated by the storm surge.

Another federal program that encourages mitigation is the Stafford Act. Enacted in 1988, the Robert T. Stafford Disaster Relief and Emergency Assistance Act created the system in which a presidential disaster declaration triggers financial and physical assistance through FEMA. The act established a cost-sharing plan of 75 percent federal funds to 25 percent state and local funds and provided for public assistance in repair, restoration, and debris removal, as well as an emphasis on mitigation with the establishment of mitigation grants.

Although various federal agencies, for example, the Department of Housing and Urban Development (HUD), the Small Business Administration, and the Environmental Protection Agency, have mitigation programs in place, FEMA is responsible for coordinating much of the disaster relief as well as disaster mitigation. FEMA has several programs that focus on mitigation, such as the Hazard Mitigation Grant Program, which provides grants to states and local governments "to implement long-term hazard mitigation measures after a major disaster declaration." This plan is intended to "reduce the loss of life and property due to natural disasters and to enable mitigation measures to be implemented during the immediate recovery from a disaster." In addition, FEMA's Pre-Disaster Mitigation Fund offers grants to state and local governments and tribal organizations for hazard mitigation planning and the implementation of mitigation projects. FEMA also administers the Flood Mitigation Assistance Program, the Severe Repetitive Flood Loss Program, and the National Flood Insurance Program.

While these programs have undoubtedly increased community resilience in some ways, they also have promoted, or failed to prevent, overdevelopment in highly vulnerable coastal areas. In fact, these programs openly act in opposition to the known sea-level rise and the need to retreat from the encroaching shoreline. They make certain that when we finally do retreat, it will be an unplanned response to major storms. Without these government subsidies, especially those granted by the Stafford Act, there would be far fewer, and far less expensive, homes along the coast. We have witnessed nearly a half-century building boom on U.S. coasts at a time when the seas are rising and are certain to rise even faster well into the future. Storm after storm can damage buildings and wipe out structures, only to have them repaired or replaced afterward with even more expensive structures, in large part courtesy of U.S. taxpayers. Frequently, storm recovery is essentially a vast urban renewal project.

In March 2014, Beth Daley of the New England Center for Investigative Reporting identified a vacation home at 48 Oceanside Drive in Scituate, Massachusetts, that has sustained significant damage from storms at least nine times in the past four decades. Each time, the owners were able to rebuild with money from the NFIP. The house was knocked off its foundation in the 1970s and raised in 2005 with $40,000 from the NFIP. In 2014, the owner sought $80,000 in federal funds to raise the home again. Donald Craig, an accountant, purchased the home in 1999 for $450,000 and was initially charged $12,000 in flood insurance premiums. But he was able to reduce the premiums nearly in half because the house was "grandfathered," based on the code and flood maps in effect when the house was constructed. Grandfathering is yet another way that we subsidize and disguise the true costs of living at the coast. For example, owners who have flood insurance when a new flood map (Flood Insurance Risk Map [FIRM]) is issued and whose structures were built in compliance with the earlier flood map are given a lower insurance rate. New England has 534 "severe repetitive loss properties," part of some 12,000 nationwide. The Union of Concerned Scientists' 2013 report, "Overwhelming Risk: Rethinking Flood Insurance in a World of Rising Seas," revealed that the NFIP has paid out nearly $9 billion to repetitive-loss properties, nearly 25 percent of all NFIP payments since 1978. Furthermore, even though repetitive-loss properties make up just 1.3 percent of all NFIP policies, they are expected to account for as much as 15 to 20 percent of future losses. In Waveland, Mississippi, vacant lots adjacent to the beach where houses had been destroyed in Hurricanes Camille in 1969 and Katrina in 2005 were for sale in 2006 for as much as $800,000. The same Union of Concerned Scientists' publication contains a summary of NFIP policy numbers, including the amount paid into the system and the value of property covered by the policies. For example, Floridians paid $1,039,000 to cover around $477,000,000 in potential obligations.

The west end of Dauphin Island, Alabama, is a very low and narrow spit that we rank as the worst development site on any American barrier island. The island has been battered by tropical storms 10 times since 1979, at a cost of approximately $80 million for an area roughly

1 square mile containing around 400 homes. Much of that money has come through the Stafford Act, which dispenses postdisaster federal funds with few strings attached. For Dauphin Island and elsewhere, these funds simply rebuild the destroyed infrastructure and houses—until the next storm takes them out again. Thanks to the Stafford Act, this $80 million has gone into repairing Dauphin Island, which adds up to more than $60,000 for each permanent resident. This does not include payments of $72.2 million made to homeowners from the NFIP since 1988, compared with only $9.3 million paid in premiums by Dauphin Island property owners over this period. There is no better example in North America of a coastal development ripe for retreat than western Dauphin Island.

The density of homes on such a narrow and low strip of sand so close to the ocean would astound a visitor who is unfamiliar with U.S. coastal construction practices, but most Americans who have spent a summer vacation at the coast have seen similar mind-boggling construction in such highly hazardous areas. Older Americans may recall a time when coastal construction consisted of "beach shacks," relatively cheap mom-and-pop vacation homes, but the twenty-first-century U.S. coast is now littered with expensive homes, hotels, and condominiums, thanks in large part to government subsidies, including the Stafford Act and the NFIP.

Older Americans may recall a time when coastal construction consisted of "beach shacks," relatively cheap mom-and-pop vacation homes, but the twenty-first-century U.S. coast is now littered with expensive homes, hotels, and condominiums, thanks in large part to government subsidies.

The Taxpayer and Emergency Spending

We should note that the funding of the Stafford Act "emergencies" is not part of the regular federal budget. Rather, this money is treated as "emergency spending," which may serve to hide the true cost to taxpayers. Reporter Kate Sheppard pointed out in 2013 that over the prior two years, disasters cost the U.S. Treasury $188 billion, nearly $2 billion a week. Hurricane Sandy was expected to cost the federal government $60 billion, and another 10 storms in the three years leading up to 2013 added up to $1 billion more in damages. Between 2000 and 2010, there were 56 declared disasters per year, compared with just 18 a year in the 1960s.

The Federal Insurance and Mitigation Administration, a FEMA program, administers the NFIP. As we explained earlier, participating communities agree to adopt and enforce ordinances that meet or exceed FEMA requirements in order to reduce the risk of flooding. This has prevented some building in flood hazard areas, but since these far-below-market-price insurance premiums were introduced, there has been more building (and more building of increasingly expensive houses) in vulnerable coastal areas than there would have been if owners had to pay insurance premiums that were not subsidized.

To discourage development on risky and environmentally sensitive areas of the coast, in 1982 Congress passed the Coastal Barrier Resource Act (CBRA, also known as COBRA), which prohibited NFIP insurance and other federal aid, such as beach replenishment and disaster relief to some designated undeveloped or lightly developed coastal barriers, like islands, spits, and mangroves that shield the mainland from wind, waves, and tides and result in unique aquatic habitats. The COBRA program explicitly states that NFIP insurance could encourage development and thus prohibited its coverage of those areas in an effort to "reduce the loss of human life, wasteful spending of federal money, and damage to fish, wildlife, and other natural resources associated with coastal barriers along the Atlantic and Gulf of Mexico coasts."

Except for a very few houses, North Topsail Beach in North Carolina is on COBRA-designated land. North Topsail Beach is an 11.1-mile-long community occupying the northern third of Topsail Island. We consider it to be the most dangerous island for development on the U.S. east coast, exceeded nationally only by Dauphin Island, Alabama. North Topsail is low and narrow (25 yards wide in some locations) and has a single escape road that easily floods and will prevent escape even in the earliest phase of a storm. The single row of dunes on the island is artificial, having been bulldozed up from the beach. The dunes, too, disappear quickly in storms. Hurricane Fran cut six inlets across the skinny island.

Seven miles of the 11-mile-long beachfront are in the COBRA program, and no better example exists of the failure of the original assumption that people would not build where there was no federal flood insurance and no federal assistance for the construction of infrastructure such as roads, beach replenishment, and storm cleanup.

The irresponsibility of development on this hazardous island is breathtaking. A number of high-rise condominiums and hotels have been built, including the high-end St. Regis Resort and Villa Capriano condo/hotels. The Topsail Reef Condo at the northern end is about to fall in along the same stretch of shoreline where a half-dozen single homes were demolished at an average cost of $30,000 in 2012. Most, but not all, of the development of North Topsail Beach took place after the COBRA designation. A 2008 article in the *Raleigh News & Observer* noted that FEMA had allocated about $6 million for public-works projects in COBRA-designated areas in North Topsail Beach, using a loophole that allowed the use of federal funds when life or property was threatened. As a result, homeowners who foolishly built in a hazardous area where federal relief funds were barred had their streets fixed, water and sewer lines repaired, and demolished homes removed, thanks to FEMA and the U.S. taxpayers. Bills to remove the COBRA designation have been introduced repeatedly in Congress but have thus far been unsuccessful.

Farther down the east coast, another COBRA-listed area is part of Kiawah Island, South Carolina, just south of Charleston. The lack of NFIP funds here, however, has apparently not stood in the way of designs to develop the extremely vulnerable southern end of the island. Kiawah Development Partners has sought to build 50 houses as well as a seawall on Captain Sam's Spit, which is the COBRA-listed area that makes up the tip of Kiawah Island. Clearly, despite the lack of NFIP coverage, developers are foolish enough to build on this very narrow, volatile strip of sand fronted by the ocean and backed by the Kiawah River. Environmental groups have fought in the courts to prevent this development, and as of the writing of this book (2015), they have been successful, but the battle is not over yet. Incidentally, the island includes an area where dolphins "strand feed"—that is, where they drive schools of fish onto the beach and actually come ashore to capture and eat them. This is a rare hunting behavior unique to the southeastern United States. Unfortunately, building in unbelievably hazardous coastal areas is not a rare behavior among developers.

The increase in coastal development in the United States has resulted in an economic crisis for both the NFIP and other insurers. Between 1991 and 2010, hurricanes and tropical storms made up 44 percent of the total number of catastrophic losses. After a particularly busy hurricane season in 2005, the NFIP had an estimated $23 billion in liabilities, compared with just $2.2 billion received in annual premiums. By 2011, the NFIP was close to $18 billion in the red (figure 2).

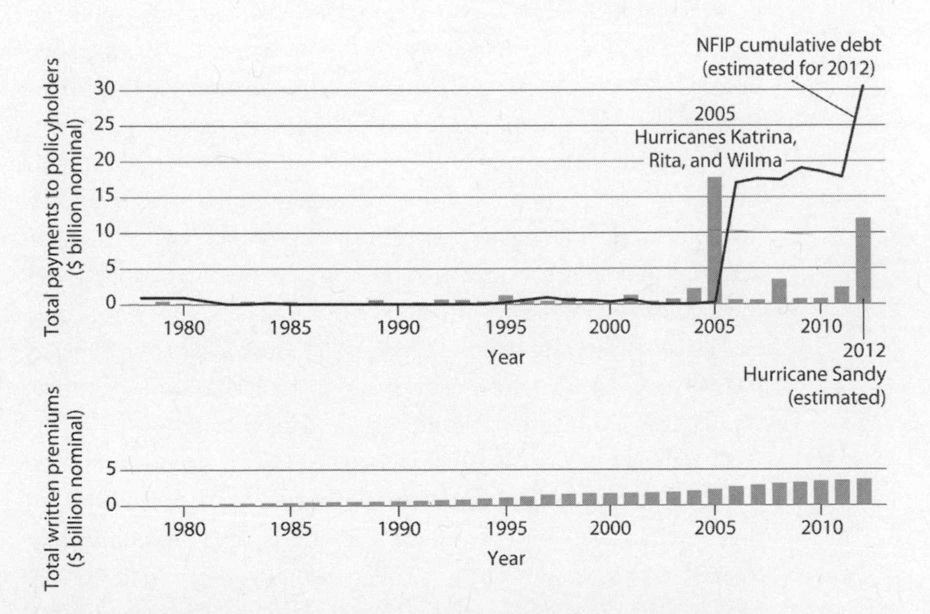

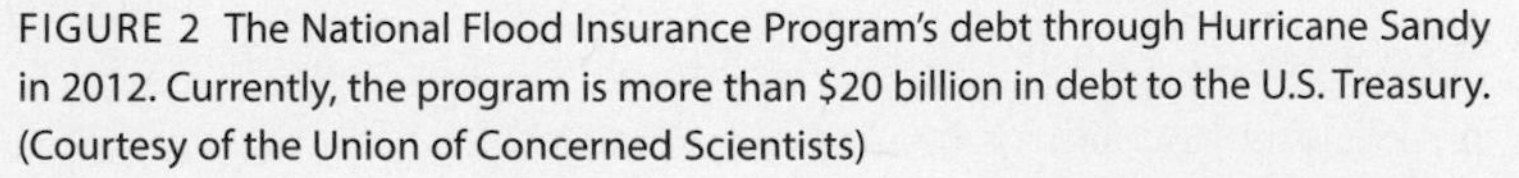

FIGURE 2 The National Flood Insurance Program's debt through Hurricane Sandy in 2012. Currently, the program is more than $20 billion in debt to the U.S. Treasury. (Courtesy of the Union of Concerned Scientists)

The Biggert-Waters Flood Insurance Reform Act of 2012 sought to address the NFIP's financial problems by phasing out the subsidized rates for newly purchased properties, lapsed policies, and new policies. This much-needed reform raised insurance premiums to reflect the true flood risks and required the updating of flood-risk maps. The elimination of the subsidized premiums is only a small step, however, in the endeavor to make this program fiscally responsible. Certainly, the NFIP has failed to deter property owners from rebuilding in hazardous areas. Unless the program is altered to offer mitigation that is more effective in preventing catastrophic storm damage and to eliminate continued payouts to homes repeatedly flooded, it will never be fiscally sound.

The Flood Insurance Reform Act of 2012 sought to end subsidized rates for 438,000 insurance policies in flood zones—chiefly for businesses, second homes, and repeatedly flooded properties. In the *Washington Post*'s *Wonkblog*, Brad Plumer reported in October 2013 that subsidies for roughly 715,000 additional properties were to be slowly rolled back as homes were sold. Unfortunately, after the widespread public uproar over the suddenly increased insurance rates, Congress is expected to enact legislation to delay the reduction in subsidies. Complicating the reform efforts are issues of economic justice: while many of the subsidies are for second homes of the wealthy, many others are for working-class families who could be forced from their homes by a significant rise in insurance rates. The numbers of homes considered to be in flood zones will dramatically increase as FEMA updates the flood maps and as the sea level rises. Just as people need to realize that we must move away from living in hazardous coastal areas prone to flooding, we also must accept that we must move away from subsidizing flood insurance.

Commentator John Stossel has long been a vocal opponent of the NFIP. He once owned an oceanfront house in Long Island, New York, that was damaged twice by minor storms before it was completely washed away. Each time, he was able to rebuild or recoup his losses, thanks to his subsidized national flood insurance. The most he ever paid in premiums was a few hundred dollars. If Stossel did not have the benefit of subsidized insurance and had to pay rates that reflected his true risk of building a house right on the beach, it is unlikely that he would have been foolish enough to repeatedly rebuild the house, and we would not have had to pay for him to rebuild it twice with our tax money.

Stossel is not the only famous wealthy homeowner who has benefited from subsidies that promote risky building on beaches. Music executive Matthew Knowles, father of singer Beyoncé Knowles and manager of her one-time band Destiny's Child, once owned a $425,000 beach home in Galveston, Texas. After Hurricane Ike damaged the house in 2008, Knowles's repairs included the construction of a concrete slab. Unfortunately, this construction ran afoul of the Texas Open Beaches Act, which ensures beach access in part by protecting "the public easement from erosion or reduction caused by development." The state sought to fine Knowles up to $2,000 per day for violation of the act. However, the American public, thanks to the Stafford Act and FEMA, along with Texas funds, came to his rescue by buying out Knowles, along with 67 other homes in the area. These are 68 homes that will never again have to be replaced or rebuilt or have their occupants rescued during a storm.

Some progress is evident as FEMA also is now funding projects that include sea-level-rise estimates through its Hazard Mitigation Grant Program. In the past, cost-benefit analysis could be based only on current and historic conditions, but now communities can use the Sea-Level Rise and Coastal Flooding Impacts Viewer, a mapping tool designed by NOAA that shows potential future inundation from daily tides along the U.S. coasts if the global sea level rises up to 6 feet. In addition, the $16 billion HUD community development block grants slated for infrastructure improvement and set to be released as the second round of the Sandy Relief Act call for grantees to provide risk assessments that include forward-looking analyses of risks to infrastructure from climate change and other hazards.

As coastal construction has increased, however, private insurers have retreated from coastal areas. Seven coastal states (Alabama, Florida, Louisiana, Mississippi, North Carolina, South Carolina, and Texas) now offer state-sponsored windstorm insurance pools because private insurers are reluctant to offer coverage in such high-risk areas. Florida's plan significantly expanded coverage when private homeowners' insurance claims following Hurricane Andrew in 1992 bankrupted 11 insurance companies and led other insurers to stop writing or renewing policies in the state. Florida's state government thus had to create the Citizens Property Insurance Corporation, a nonprofit insurer of last resort. Since this entity was created, more and more private insurers have left the market. Now Citizens Property is the state's largest property insurer, covering more than 1.5 million policyholders. Some Florida legislators fear that a bad hurricane season could threaten the state's solvency in trying to ensure the payout of claims. A similar state-formed nonprofit insurer of last resort was started in Louisiana in 2004, and it currently runs a $56 million deficit caused by costs from Hurricanes Katrina, Rita, and Isaac.

The experiences of the NFIP, as well as those of the private insurance industry and state-created insurance programs, show that we cannot continue business as usual. We have created the potential for greater "natural" disasters by subsidizing construction in areas that are highly vulnerable to damage from flooding and winds from storms. We have created a situation in which relatively mild storms can result in widespread financial devastation, even putting states like Florida and Louisiana at the risk of insolvency. Moreover, the situation will only worsen with sea-level rise, as we can expect increased damage from storms and, quite likely, increasingly powerful storms due to changes in the climate, especially the warming of the oceans' surface waters.

We have created a situation in which relatively mild storms can result in widespread financial devastation, even putting states like Florida and Louisiana at the risk of insolvency.

Along with insurance support, beach replenishment is yet another way in which governments subsidize coastal building and rebuilding. Andy Coburn of the Center for the Study of Developed Shorelines at Western Carolina University estimates that over the past decade, the U.S. Army Corps of Engineers spent between $125 million and

$150 million per year, and FEMA spent an additional $25 to $30 million per year, on beach replenishment. Coburn also estimates that the Sandy Relief Act authorizes approximately $3.5 billion for beach replenishment and that roughly $7 billion has been spent on sand projects nationwide, with about one-quarter of the spending on projects based in Florida.

Over the years, the cost of replenishing beaches has shifted from the state and local levels to the federal level, and this is in part thanks to the Stafford Act. Federal taxes account for approximately three-quarters of the spending on replacing beach sands, with state and local governments and a small share from private landowners covering the rest. By deeming beach replenishment as infrastructure repair, the Stafford Act effectively sends the bill for it to the federal government. Whereas the local and state governments initially paid for beach replenishment, now the states and seaside communities with an engineered replenishment merely wait for a damaging storm, and once the president declares a disaster, federal funds begin pumping sand back onto the beach. A big storm saves communities lots of money because they are assured of continuing replenishments for the foreseeable future.

The National Flood Insurance Program subsidizes the private homeowners and encourages building in areas where homes will have to be repeatedly repaired or rebuilt. The Stafford Act similarly subsidizes risky infrastructure construction on a communitywide level. The Stafford Act is designed to respond mainly to sudden natural disasters, with the one exception being aid for droughts. It is not designed to respond to gradual changes in sea level. For instance, coastal erosion, a gradual environmental process, is not covered by the Stafford Act, yet it threatens many developed coastal areas. Unfortunately, when disaster strikes, the funds are released, and in the rush to repair communities, much of the damaged or destroyed infrastructure is simply rebuilt in the same flawed fashion to await destruction by the next major storm.

Some FEMA programs have the potential to deal with challenges associated with sea-level rise. In particular, one form of mitigation—property acquisition programs—has repeatedly proved to be a wise investment of federal funds by encouraging retreat from hazardous areas. Following the 1993 flooding of the Mississippi and Missouri Rivers in the Midwest, legislation reforms known as the Volkmer Amendment shifted the cost-sharing for property acquisition from 50–50 to a federal share of 75 percent, compared with the state and local share of 25 percent. Missouri took advantage of the buyout program to remove 2,000 families from the floodplain by the time historic floods struck again in 1995. Even though in some areas, the 1995 flood was the third largest in history, it resulted in much less damage than that from the 1993 floods. An estimated 95 percent of the properties in Missouri bought out after the 1993 floods also would have been inundated by the 1995 floods. Following the 1993 floods, in St. Charles County, Missouri, there were 4,227 applications for supplemental federal assistance, while there were only 33 applications in 1995. FEMA's program expenditures for the Missouri flooding dropped from a whopping $26 million in 1993 to less than $300,000 in 1995.

FEMA's Hazard Mitigation Grant Program makes buyouts after disasters, but three other FEMA programs offer buyouts that do not depend on disasters: the Flood Mitigation Assistance Program, the Pre-Disaster Mitigation Program, and the Severe Repetitive Loss Program. Funding for these programs will need to be significantly bolstered to meet the challenges of sea-level rise, but the acquisition of property offers the most effective permanent solution to prevent the catastrophic loss of life and property. Fran McCarthy observed that the city of Kinston, North Carolina, actively sought to move neighborhoods to preserve school districts and social communities in part of an acquisition of 1,000 properties in 1999, in response to repeated flooding from hurricanes. As seen in the New York post-Sandy buyout programs, incentives can be offered to keep communities intact, as this program does by relocating homeowners in the same counties in order to maintain community and to sustain the tax base. This, however, will be a significant challenge in low-lying coastal areas as the seas continue to rise. In some cases, it will be impossible to relocate communities locally. For instance, in North Carolina, the elevation of the coastal plain behind the shoreline is so low that it would make no sense to try to relocate people to another section of the county, as that land will inevitably be inundated as well. The situation is even worse in Florida (see chapter 3).

Acquiring properties and removing houses and infrastructure make particular economic sense in an age of sea-level rise. If we do not remove communities from low-lying coastal areas, taxpayers will continue to be forced to bail out those affected by "natural disasters."

The time has passed for U.S. taxpayers to be required to subsidize coastal living. The libertarian R Street Institute, using Government Accounting Office information, reports that 78.8 percent of subsidized policies are in counties that rank in the top 30 percent of home values, while fewer than 1 percent are in counties that rank in the bottom 30 percent. Many of the subsidies are for second homes of wealthy homeowners. We subsidize their flood insurance, and when their houses are destroyed, we provide emergency relief funding to rebuild their homes. This welfare for the rich must stop. In an age of climate change with rising seas and the increased severity of storms, we can no longer afford to hide the true costs of building in hazardous coastal areas. We must recognize that we do not owe anything to anyone foolish enough to build in such obviously dangerous places. Rather, we should penalize such behavior. Our tax money could be better spent encouraging and facilitating a planned and managed retreat from the coast—relocating people to higher ground instead of repeatedly rebuilding in many areas that we never should have built in the first place.

PLATE 1 (*top*) The concrete slab is all that remains of the Pilkey homesite (the authors' parents and grandparents) at Waveland, Mississippi, after Hurricane Katrina passed through in 2005. In 1969, Hurricane Camille flooded, but did not destroy, the house, which was five blocks from the beach. Hurricanes Camille and Katrina inspired the writing of this book. (Photograph by Orrin H. Pilkey)

PLATE 2 A storm wave striking the seawall in front of a development immediately adjacent to the shoreline in Saint-Malo, France. The tidal amplitude in this area is very large, and if this storm had struck at low tide, the wave activity on the seawall would have been less spectacular. (Photograph used by permission of Patrick Manac'h)

PLATE 3 (*top*) Apparently the community of the aptly named Washaway Beach, Washington, wants to make very sure that no one will walk or drive down this road to this rapidly eroding beach. Three rows of small, low-cost cottages have fallen into the ocean here, where little effort has been made to stabilize the shoreline. Letting buildings fall into the sea is a form of retreat from the shoreline. (Photograph used by permission of Norma Longo)

PLATE 4 A rendering of a $2 billion development called the Worldcenter shows an example of the construction boom under way in Miami, Florida. The intensity of development in this city so obviously threatened by sea-level rise is stunning. (Photograph used by permission of Miami Worldcenter Associates)

PLATE 5 (*top*) A flooded New Orleans after Hurricane Katrina passed through in 2005. Ironically, the baseball field in the background is at a higher elevation than most of the surrounding houses. This type of post-storm scene will become increasingly common as sea-level rise progresses. (Photograph courtesy of Federal Emergency Management Agency [FEMA] / Liz Roll)

PLATE 6 The dampening effect of salt marshes on storm surges is assumed to be a major factor in protecting New Orleans from future flooding. A view of Pecan Island, Louisiana, showing the extensive marshes that make up much of the Mississippi Delta. Laced throughout these marshes are canals carved by dredges to provide access to oil-drilling platforms. Obviously, the canals have contributed to the loss of marshland. (Photograph used by permission of Airphoto / Jim Wark)

PLATE 7 (*top*) Some of the citizens of Far Rockaway Beach on the southern shore of Long Island, New York, are confident that they can beat the sea-level rise. The chances are very good that they are wrong. The sign was painted shortly after Hurricane Sandy in 2012. (Photograph courtesy of Leonard Zhukovsky [/gallery-1024723p1.html/] Shutterstock.com)

PLATE 8 This massive beach-replenishment / seawall project coupled with dune construction in Scheveningen, Netherlands, is typical of such projects in that low-lying country. Because the Dutch live on a delta with little high ground, they depend on coastal engineering for their very survival. (Photograph used by permission of J. Andrew G. Cooper)

PLATE 9 (*top*) After the 1900 hurricane that killed at least 6,000 people in Galveston, Texas, the elevation of the entire small town was raised by pumping sand onto the island. Here, engineers are preparing to raise a fire hydrant to the new ground surface level. (Photograph courtesy of the Rosenberg Library, Galveston, Texas)

PLATE 10 A low-cost seawall shown at spring tide at a small coastal fishing village near Buenaventura, Colombia. For developing countries, substantial seawalls are often too expensive, so walls like this one, made up of local materials, are used. (Photograph used by permission of Bob Morton)

PLATE 11 (*top*) The skyline of Boston, Massachusetts, a city highly vulnerable to sea-level rise. Most of the downtown was constructed by filling in the bay. (Photograph used by permission of Tim Grafft, Massachusetts Office of Travel and Tourism [MOTT])

PLATE 12 Clearly, a storm surge here on this low-lying neighborhood (Makoko) in the lagoon of Lagos, Nigeria, will do substantial damage. Sea-level rise will undoubtedly force these homes to be moved very soon. (Photograph courtesy of AP Photo / Sunday Alamba)

PLATE 13 (*top*) A conceptual drawing of a planned seawall and reservoir system for the city of Jakarta. The design resembles the Garuda, from a Hindu myth about a powerful bird-god. (Image courtesy of Consortium NCICD / Design: KuiperCompagnons)

PLATE 14 Benidorm, Spain, was a sleepy fishing village into the 1950s but is now a major tourist site lined with high-rises, which will make the eventual retreat from the shoreline both expensive and complicated. (Photograph used by permission of Norma Longo)

PLATE 15 (*top, opposite page*) Here, at the northern end of North Topsail Beach, North Carolina, one row of houses has already been condemned and removed, and now the second row is threatened. Multiple beach-replenishment projects have been carried out here, the latest of which lasted only six months. In 2015, a large sandbag seawall was placed in front of the buildings, including in front of the Topsail Reef condominiums at the top of this photo. It is clear that nature is winning here but also that city officials are fighting it all the way and rejecting the notion of retreat. (Photograph used by permission of Andy Coburn, Program for the Study of Developed Shorelines, Western Carolina University)

PLATE 16 (*bottom, opposite page*) This flooding of the low-lying western end of Dauphin Island, Alabama, is a result of a combination of the October 27, 2015, king tide and rains from distant Hurricane Patricia. The island has suffered severe property damage five times, starting with Hurricane Frederic in 1979. Instead of the inhabitants sensibly retreating from the shore, the shoreline has been held in place repeatedly by replenishment, and plans are afoot for additional engineering to preserve the beachfront buildings. There is no better example of an island segment where development should be completely abandoned—houses moved or demolished and the island left to the seagulls. (Photograph used by permission of Sam St. John, www.flythecoast.com. Copyright 2015)

PLATE 17 (*above*) The Inupiat community of Kivalina, Alaska, along the Chukchi Sea, is completely surrounded by a seawall. Due to a combination of melting permafrost, sea-level rise, and the late arrival of snow and ice in the fall, the village is now in serious danger from erosion. It is one of 12 Alaskan coastal subsistence villages that are endangered. Serious consideration is being given to moving these communities to the mainland, albeit at great expense. (Photograph courtesy of AP Photo / Northwest Arctic Borough via the *Anchorage Daily News*)

PLATE 18 (*top*) Majuro Atoll is the capital of the Marshall Islands in the Pacific Ocean. Atoll communities, often viewed as canaries in a coal mine, with a total worldwide population of around 1 million, are in critical danger from inundation by the rise in sea level. Plans are afoot to move atoll inhabitants to other countries such as New Zealand, Papua New Guinea, and Sri Lanka. (Photograph used by permission of Greg Vaughn)

PLATE 19 The Gold Coast, which Australians call their Miami Beach. The Gold Coast has the same problem as much of the high-rise-lined coast of Florida—that is, what to do when the sea level rises 2 or 3 feet above its current level. We would expect, as will likely happen in Florida, that initially the beach will be replenished, to be followed by the construction of large seawalls built on both sides of this barrier spit. (Photograph used by permission of J. Andrew G. Cooper)

PLATE 20 (*top*) Aerial photograph showing the impact of a storm surge on La Faute-sur-Mer and L'Aiguillon-sur-Mer in Vendée, France. The devastating storm Xynthia in 2010 caused the flood shown here. (Photograph courtesy of MAXppp / Landov)

PLATE 21 Aerial photograph of the same area in Vendée, France, shown in 2011, one year after the storm Xynthia wreaked havoc on the area and killed 29 people. (Photograph courtesy of MAXppp / Landov)

BIENVENIDOS
HOTEL
MIRADOR DEL PONIENTE

PLATE 22 (*top, opposite page*) The Hotel Mirador del Poniente was abandoned because of the rapid local rise in sea level (caused by sinking of the land related to an earthquake) on Isla Soldado on the Pacific coast of Colombia. In poor communities like this one, money to engineer the shoreline to protect buildings is rarely available. As a consequence, retreat from the eroding shoreline is a continual process, by moving buildings, abandoning them, demolishing them, or letting them fall into the sea. (Photograph used by permission of William J. Neal)

PLATE 23 (*bottom, opposite page*) This elevated concrete building is a cyclone shelter in Chittagong, Bangladesh. These shelters are found well inland throughout this low-lying delta. As flooding increases, the delta inhabitants will eventually become environmental refugees and flee the area, perhaps even across the border to Myanmar or India. (Photograph courtesy of AP Photo / A. M. Ahad)

PLATE 24 (*above*) The Great Wall of India on a foggy day. This double fence is intended to prevent the migration of a great number of Bangladesh citizens who are expected to attempt to flee to India to escape rising seas and disappearing land on the Ganges Delta. (Photograph courtesy of AP Photo / Ramakanta Dey)

PLATE 25 (*above*) This house in Highlands, New Jersey, is being raised following Hurricane Sandy in order to comply with the Federal Emergency Management Agency's height regulations to enable the owner to receive federal flood insurance. This is an example of adaptation: a temporary alternative solution to the problem of rising seas that cause higher storm surges. (Photograph courtesy of the U.S. Geological Survey [USGS] / Rosanna Arias)

PLATE 26 (*top, opposite page*) Imagine the future of these Clearwater, Florida, high-rises on the Gulf coast of Florida in the next hurricane or in coming decades as the sea level rises. Is it likely these buildings, and hundreds more like them, can be moved back from the shoreline? (Photograph used by permission of Andy Coburn, Program for the Study of Developed Shorelines, Western Carolina University)

PLATE 27 (*bottom, opposite page*) This line of high-rises along the shoreline of Recife, Brazil, has been protected by the construction of a boulder revetment. The result of this protection has been the loss of the beach, which was the reason for the community's existence in the first place. The narrow band of sand at the top of the wall is meant to replace the natural beach. The next step as the sea level rises will be to increase the height of the revetment. (Photograph used by permission of J. Andrew G. Cooper)

PLATE 28 (*top*) The debris along this stretch of England's Holderness Coast is a mixture of World War II pillboxes and buildings from a destroyed village. Since the Roman invasion 2,000 years ago, 28 villages have gone into the sea and now reside on the continental shelf. (Photograph used by permission of Neil A. White)

PLATE 29 The eroding bluff along England's Holderness Coast continues to threaten houses and infrastructure. The shoreline here has eroded approximately 2 to 3 miles over the past 2,000 years, with very little contribution from sea-level rise. The lesson of the Holderness Coast is that the problems affecting developed shorelines will likely continue for many generations to come, a fact that should be considered in coastal planning. (Photograph used by permission of Neil A. White)

7

Coastal Calamities

How Geology Affects the Fate of the Shoreline

It is an absolute and inescapable fact that humans will move back from the rising sea levels. Just how they will make such moves will vary globally according to differences in coastal cultures, national wealth and politics, and the physical environments of the shorelines. Likewise, the timing and orderliness (whether we plan to step back or flee in disarray) of retreat will vary around the world. Bangladesh (as discussed in chapter 5) is already beginning a retreat on a small scale. This nation has little engineering tradition and lacks the economic resources even to attempt to out-engineer Nature. At the other end of the spectrum, the Dutch, who live on the combined Rhine-Meuse-Scheldt river deltas, may well be the last low-lying nation to retreat because they have a strong engineering tradition, have enough financial resources and national support to attempt engineering solutions, and, most important, have nowhere else to go.

The coastal environment varies by region. California has a narrow continental shelf and consequently has storm surges of only a few feet. This may be why the homes of the wealthy in Malibu are unwisely permitted to be built right next to the beach. In contrast, the continental shelf off Mississippi is very wide, and the topography is very low. Storm surges as high as 28 feet have occurred twice in the past 50 years in Waveland and Bay St. Louis, Mississippi. Therefore, retreat will likely occur much earlier in Mississippi than in Malibu.

The future of village dwellers on coral atolls in the Pacific and Indian Oceans largely depends on the generosity of other countries to provide funding and a place to resettle, but in any case, the villagers absolutely must relocate soon. Arctic Inupiat villages along the Bering and Chukchi Seas and Alaska's North Slope also must retreat soon. These villagers have the good fortune to be part of a wealthy society, even though the villages themselves are quite poor. To maintain their subsistence culture, their retreat will be very costly because of the expense to relocate and rebuild infrastructure in extreme weather and soil permafrost conditions. The Inupiat villages in Siberia are likely to be less lucky, and their culture and subsistence existence may disappear because the Russian government has a different outlook in this regard.

The barrier islands in rich countries like the United States and Spain are being held in place by seawalls or beach replenishment. Ultimately, in a time of rising sea level, all such communities held in place against nature's wishes will be walled, and the beaches will eventually disappear. Less-developed countries have little money to hold the shoreline in place, so villagers there must retreat, often moving their buildings back apace with erosion. In the long term, because they have no choice but to move inland, the future quality of beaches (except for pollution) is likely to be best in these countries.

Barrier islands may be narrow or wide. Residents of the narrow ones—like the Outer Banks of North Carolina; Santa Rosa Island, Florida; and western Dauphin Island, Alabama, which are often only 100 or 200 yards wide—will need to retreat much sooner than will the residents of the Sea Islands (e.g., Hilton Head and Jekyll Island) of Georgia and South Carolina, which are higher in elevation (because of a greater volume of dune sand), may be as much as 5 miles wide, have few hurricanes, and usually are subjected to low waves. On narrow or wide islands, buildings built right next to the beach will soon have to be moved, or will fall into the sea, or will be seawalled.

Storm Surges, Nature's Greatest Threat

A *storm surge* is a short-term abnormal rise of water caused by a storm, beyond the typical tide (table 4). Riding ashore on the top of the rising

TABLE 4
"Notable Surge Events" in the United States

Event	Surge
Hurricane Sandy, New York, 2012	13 feet
Hurricane Ike, Texas, 2008	15 to 20 feet
Hurricane Katrina, Mississippi, 2005	25 to 28 feet
Hurricane Dennis, West Florida, 2005	7 to 9 feet
Hurricane Isabel, Chesapeake Bay, 2003	8 feet
Hurricane Opal, West Florida, 1995	24 feet
Hurricane Hugo, South Carolina, 1989	20 feet
Hurricane Camille, Mississippi, 1969	24 feet
Hurricane Audrey, Louisiana, 1957	12 feet
New England Hurricane, Long Island and southern New England, 1938	10 to 12 feet
Galveston Hurricane, Texas, 1900	8 to 15 feet

Source: Data from National Weather Service, National Hurricane Center, "Storm Surge Overview," NOAA, http://www.nhc.noaa.gov/surge/.

waters are the waves that do a lot of the physical damage to buildings in a surge. Flooding is another damaging aspect of storm surges.

Flowing water is a powerful, destructive force. Rapidly moving floodwaters scour sediment and transport debris, which crashes into everything in its path, including buildings, utilities, and vehicles. The surge then deposits this debris inland, blocking roads and other structures.

Unlike shoreline erosion, which is usually a continuing, gradual event, storm surges are quick and brief catastrophes. If shoreline erosion is a problem in a big storm, it is usually restricted to the first row or two of houses, but a storm surge can cause major damage several miles inland. The potential for destruction from a storm surge varies widely from coast to coast and is one of the most important reasons that development should be moving back from the coast.

> Like the common real-estate mantra, "location, location, location," the three most important factors affecting storm-surge vulnerability are "elevation, elevation, elevation."

As all things in nature seem to be, a storm surge is complex and is affected by the following: the intensity and direction of the storm winds; the forward speed of the storm; the lower atmospheric pressure in storms; the size of the storm; the storm's angle of approach to the shoreline; the local tidal stage; the shape of the shoreline (especially the presence of bays and estuaries, which funnel and can increase surges); the inlet's size and location on barrier-island coasts; and, most fundamentally perhaps, the width and slope of the continental shelf. Like the common real-estate mantra, "location, location, location," the three most important factors affecting storm-surge vulnerability are "elevation, elevation, elevation."

We can perceive the effect of the continental shelf width on a storm surge by assuming that a storm is approaching a shoreline perpendicular to the coastline. The depth of the water at the edge of the continental shelf along much of the southeast United States and the mid-Atlantic coast is around 300 feet. Thus two storms of the same magnitude heading for the shore would push the same amount of water onto the continental shelf. If our hypothetical storm hits Miami, which has a narrow (5 miles) and steep shelf, there is lots of vertical space offshore to accommodate the storm-driven water. In contrast, because the wide (80 miles) shelf off Georgia is flat and must accommodate the water in a smaller vertical space near the shoreline, the storm-surge overflow onto the land will be much larger than that of Miami.

During a storm surge, knowing the stage of the tide is important because a storm striking at high tide could cause a much larger surge and more damage than one that strikes at low tide. And, of course, as the sea level continues to rise in the future, the height of future surges will increase proportionately. As an extreme example, if a big storm at low tide were to strike the Bay of Fundy, which separates New Brunswick and Nova Scotia, the storm surge might have negligible impact. This is because the tide range, the highest in the world, is around 50 feet. Conversely, if the storm struck at high tide, and especially during a high spring tide, the inundation on land would be most impressive.

According to NOAA's storm-surge models, the maximum theoretical storm surge from a monstrous hurricane would be 33 feet each for New York's JFK airport, the Big Bend Gulf Coast in Florida's Panhandle, and the shoreline reach north of Myrtle Beach, South Carolina.

The greatest potential storm surge from the monstrous modeled hurricane in all of the United States is in New Bedford, Massachusetts, with a potential surge of 38.5 feet above the expected tide level. The reason for this great height is that the city is at the end of a narrow estuary that will create a funneling effect on the surge.

Table 5, modified from the 2013 "CoreLogic Storm Surge Report," identifies the number of buildings threatened by storm surge in various states. The table is based on a worst-case scenario of the right-front quadrant of a storm striking the coast head-on at high tide. Florida clearly has the largest number of buildings at risk, partly because of the length of its shoreline (1,350 miles). In addition to shoreline length, population trends and geographic/geologic factors were considered in determining the risk status.

The loss of life from storm surges can be extreme. The deadliest storm surge accompanied Cyclone Bhola, which struck portions of India and Pakistan and what is now Bangladesh on the Ganges Delta. Perhaps as many as 500,000 people died in that 1970 storm. Cyclone Nargis killed 138,000 people in Myanmar (Burma) in 2008, a death toll increased by the military dictatorship's reluctance to allow international ships just offshore to help the survivors. The combined

TABLE 5
Number of Properties at Risk from Storm Surges in Various Coastal States

Rank	**State**	**Number of Properties at Risk**
1	Florida	1,489,000
2	Louisiana	411,000
3	Texas	370,000
4	New Jersey	360,000
5	Virginia	329,000
6	New York	280,000
7	North Carolina	232,000
8	South Carolina	197,000
9	Georgia	118,000
10	Massachusetts	107,000

Source: Howard Botts, Wei Du, Thomas Jeffery, Steven Kolk, Zachary Pennycook, and Logan Surh, "Corelogic Storm Surge Report," 2013, http://www.corelogic.com/research/storm-surge/corelogic-2013-storm-surge-report.pdf.

wind and storm surge of Typhoon Haiyan in 2013 killed at least 5,000 people in the Philippines. The wind, waves, and 15-foot storm surge from a Category 4 hurricane that struck Galveston, Texas (on a barrier island), in 1900 killed somewhere between 6,000 and 12,000 people. The estimate of dead from the 1900 hurricane varies because no attempt was made to accurately count casualties on the mainland. Beginning with the 2015 hurricane season, NOAA's National Hurricane Center is using an experimental storm-surge graphic to warn residents on the Gulf and Atlantic coasts of the potential storm surge from an impending storm.

The Atoll Nations: The Canaries in the Coal Mine

Atolls are coral islands, whose origin Charles Darwin famously discerned in 1835. The coral reefs once formed a ring around volcanic peaks that extended above the ocean surface. As the seafloor spread, the volcanoes moved away from the mid-ocean ridge into deeper water to become submerged. The reefs, however, continued building upward and maintained the ring shape that characterizes the atolls of today. These ring-shaped islands surround an interior lagoon. Darwin got it all right, except he didn't know about the phenomenon of seafloor spreading, which was discovered 150 years after his participation in the voyage of the *Beagle*.

Because the offshore slope (the sides of the volcano) is steep, storm surges there are small. But it takes only a small storm surge to flood a low-lying atoll. Typically on the open-ocean side of the atoll's rim is a continuous ridge 15 to 20 feet high, which is made up of coral reef fragments and debris tossed up in storms. The ridge usually is not habitable because it is frequently overwashed by storms. Rather, the living space is a flat area behind the ridge, typically 6 feet or so above sea level.

Table 6 lists some of the atoll island nations and communities in the Pacific Ocean. The one exception is the Maldives atolls, which are in the Indian Ocean. All the numbers are approximate, and the number of islands often includes small, uninhabited islands. The populations indicate the number of environmental refugees who soon will need to be relocated in one way or another. In total, more than a

TABLE 6
Atoll Nations and Territories in the Pacific Ocean

Name	Population	Number of Islands	Status
Cook	66,000	15	Independent
French Polynesia	263,000	24	Territory of France
Kiribati	103,000	34	Independent
Maldives	270,000	26	Independent
Marshall Islands	68,000	24	Independent
Micronesia	106,000	607	Independent
Palau	20,000	200	Independent
Tokelau	1,170	3	Territory of New Zealand
Tuvalu	10,000	9	Independent

million people live on atolls or other low oceanic islands in both the Indian and Pacific Oceans.

Some of the atoll nations listed as independent have a semiformal relationship with a larger country. For example, the Marshall Islands have a "free association" with the United States, which provides defense and social services plus grants for infrastructure assistance. New Zealand also has special relationships with some of the island nations. These free-association relationships are likely to be a factor in funding future relocation.

Perhaps the biggest single problem currently facing the people on these islands is the precarious source of acceptable drinking water. The freshwater lens under the islands floats on top of seawater in the reef-rock substrate. As the sea level rises, the freshwater lens thins and, in some places, has disappeared altogether. That is, the contact or interface between freshwater and saltwater moves upward as the sea level rises, contaminating wells. Not only is the shrinking lens problematic, but pollution of this groundwater from surface contamination infiltration is a frequent occurrence. An alternative source of drinking water is the collection of rainwater, but the larger towns cannot depend on this source. In addition, the salty groundwater leads to the destruction of food crops like coconuts and the root crops taro and pulaka. In some communities, people have taken up planting in old oil drums because of the poor soil conditions caused by the intrusion of saltwater.

The growing number of unsanitary conditions on these small, shrinking islands, many with densely packed populations, is adding to the problem. Some visitors report very poor sanitary conditions throughout and around densely populated villages where trash and piles of dog, pig, and human excrement abound. The problem is especially serious in the larger towns like Majuro in the Marshalls and Tarawa in Kiribati (more in chapter 9).

The atolls' solution to sea-level rise must be more than retreat. It will be complete abandonment and complete relocation, and not merely to another atoll. Relocation will mean moving body and soul to new lands, usually the mainland, with strange vegetation, unfamiliar foods, and unfamiliar customs. The islanders' future will not be easy and brings into serious question the future existence of island governments and culture.

The atolls' solution to sea-level rise must be more than retreat. It will be complete abandonment and complete relocation, and not merely to another atoll.

Arctic Shorelines

The Arctic shorelines in Siberia, the United States, and northern Canada have been hit by a triple whammy. The sea level is rising at rates not unlike the rates at lower latitudes, except in parts of Scandinavia, where the land is rebounding from the removal of the weight of massive glaciers. This uplift leads, of course, to a lowering of the local sea level. Because the massive continental glaciers never reached the shorelines of the North Slope of Alaska or the adjacent coasts of northern Canada and Siberia, fluctuations in the elevation of land are probably not important factors in sea-level rise for these areas.

What is important is the Arctic's rapidly warming climate, which is leading to the loss of permafrost in beach sand and coastal soils, plus changes in the annual cycle of sea ice. The loss of ice cementing the sediment (permafrost) in beaches and bluffs increases the rate of erosion by wave action. Simultaneously, the wave activity on Arctic beaches is increasing because the seasonal sea ice is disappearing earlier and is refreezing later. In the Chukchi Sea, for example, sea ice used to form along the shoreline in September, well before the November storm season arrived. Consequently, by the time the storms did arrive, the shoreline was held fast in ice. Now the protective ice does not form until after the storms have begun in November, to the detriment of the exposed, ice-free beaches.

More than a dozen Alaskan Inupiat villages along Arctic shores are threatened by flooding and the retreating shoreline. The population of most villages is only between 200 and 500 because a larger population would make a subsistence existence impossible. All these villages are seriously considering a managed retreat, and in fact they have little choice. The problem is that the cost of moving tiny villages in the Arctic is prohibitive. The U.S. government requires (and the inhabitants desire) heating, dependable electricity, running water, sewers, and garbage disposal. Complicating the situation is the high cost of any construction in remote permafrost regions, especially those where the permafrost is melting, which causes roads and buildings to sink.

Table 7 lists Inupiat villages along the Chukchi Sea and open Arctic Ocean along the North Slope of Alaska. Similar waterfront villages can be found in Siberia and northern Canada. The number of potential environmental refugees here, however, is minute compared with the

TABLE 7
Inupiat Villages on the North Slope, Alaska

Name	Population
Atqasuk	233
Barrow	4,212
Kaktovik	293
Kivalina	374
Kotzebue	3,201
Point Hope	674
Point Lay	247
Shishmaref	580
Wainwright	556

number of refugees from the midocean atoll islands, a few thousand versus a million. Yet the cost of retreating from the shoreline in the Arctic is immense, almost an economic impossibility. So why don't the Inupiat move to the nearby "big cities," such as Nome or Kotzebue or even Anchorage? The answer is twofold. Moving to town would be the end of their subsistence lifestyle because too many hunters and fishers would quickly wipe out the local food source. Great skill in subsistence hunting and fishing is of little value in an urban environment. The second problem is the Inupiats' tendency toward alcoholism, which could not be dependably controlled in larger towns. This problem is the reason that most Alaskan villages prohibit the possession of alcohol.

Shishmaref is a village situated on a barrier island in the Chukchi Sea, just below the Arctic Circle. The U.S. Army Corps of Engineers has estimated the cost of moving the village to the mainland behind the current town site to be around $300,000 to $400,000 per village inhabitant. Such an expenditure to move a small Arctic village could never achieve the favorable cost-benefit ratio normally required for corps projects. But because subsistence Inupiat villages are a rare remaining fragment of the past and much respected in America, normal cost-benefit requirements don't apply. Already, the cost to the federal government of the seawalls built in front of Shishmaref has probably exceeded the value of all the buildings in the village. To date, little progress in the actual moving of villages has been made.

The Lowlands

River deltas are the world's lowest elevation, extensively developed coastal areas. The most densely developed of all is the Ganges–Brahmaputra Delta, which is almost entirely occupied by the country of Bangladesh, with a corner belonging to West Bengal, India. Bangladesh is a poor country that likely will not be able to hold back sea-level rise in any fashion. Where do the people of this—the most densely populated country in the world—move as the sea level rises? There may be some room to spare in bordering Myanmar and India, but international boundaries probably will prevent mass legal migrations of environmental refugees.

Myanmar's Irrawaddy Delta is densely packed with people and has made almost no preparations for sea-level rise and storm surges. Hence the devastating Cyclone Nargis in 2008 killed more than 138,000 people. At least Bangladesh has organized warning systems and built numerous elevated concrete shelters to which the local people can flee during storm surges. Although these relatively low-tech structures allow people to wait out flood events, they are, at best, very temporary and unreliable solutions to the threat of sea-level rise.

In contrast to these deltas occupied by poor countries is the Rhine Delta in the Netherlands. The event that precipitated the Netherlands' obsession with sea-level rise and storm protection was a northwesterly storm in 1953 that struck at spring tide and produced a storm surge of more than 18 feet, killing 1,836 people (see chapter 4). The storm was much like the 1962 Ash Wednesday nor'easter on the U.S. east coast, which struck during a high spring tide and moved away very slowly through five high tides. One of the most powerful storms to hit the east coast of the United States, it resulted in 40 fatalities.

Other deltas of note include the oil-rich Niger Delta in Nigeria and the Nile Delta in Egypt, both of which have been greatly affected by the loss of sediment from dammed rivers and canal construction. The largest river without a single dam is the Fly River in Papua New Guinea, with a delta on the Gulf of Papua. Interestingly, the Amazon River does not form a true delta. Instead, its sediment load, which is mostly mud rather than sand, is deposited in deep water far offshore. (The large Mississippi Delta in the United States was discussed along with New Orleans in chapter 3.)

River deltas rank just behind the atoll nations and the Arctic Inupiat villages in the level of urgency needed for planning and carrying out an immediate response to the rise in sea level. Undoubtedly, retreat is in store for most delta residents and atoll dwellers alike.

River deltas rank just behind the atoll nations and the Arctic Inupiat villages in the level of urgency needed for planning and carrying out an immediate response to the rise in sea level. Undoubtedly, retreat is in store for most delta residents and atoll dwellers alike.

Other flat, low-lying stretches of land next to the sea include the northeastern corner of mainland North Carolina, which is one of three areas in the lower 48 states most threatened by sea-level rise. The Mississippi Delta and South Florida are the other two areas with this distinction.

The mainland behind the Outer Banks of North Carolina is a broad, flat, swampy area (including the Great Dismal Swamp) surrounding Pamlico and Albemarle Sounds. Compared with the residents of the barrier islands facing the sea, the mainland folks are, on average, less prosperous, less visible, less educated, and less influential in the North Carolina political scene. Probably more than 100 small towns, some consisting of a few houses and usually a church, are located there, and many have an elevation of less than 5 feet. Some of the larger towns include Manteo (population 1,340), Manns Harbor (821), Elizabeth City (18,470), Swan Quarter (324), Bath (247), Aurora (520), Washington (9,744), Columbia (863), and Plymouth (4,107).

In 2011, the Category 1 Hurricane Irene affected much of the area, flooding the lower portions of most of these towns and inundating almost all of Manteo. As usual, the news media in North Carolina concentrated on Irene's damage to the buildings on the barrier islands on the Outer Banks and almost ignored the detrimental effects on the mainland villages.

In 2010, the Federal Emergency Management Agency (FEMA) gave North Carolina a $5 million grant to map the northeastern corner of the state, including maps of the predicted storm-surge levels with various sea-level rise scenarios of 1.5 feet, 3 feet, and 5 feet likely to occur by the year 2100. This kind of information is critical to community planners, to individuals building or buying houses with the hope of eventually leaving them to their children, and also to industry scouts seeking new locations for businesses and factories.

Unfortunately, fearing the impact on real-estate prices and local economies, the state government prohibited the publication of the storm-surge maps, which now sit in a drawer in a cabinet in someone's office. This reckless, irresponsible act by the state government lost it a critical opportunity to begin considering its options, including planning a gradual retreat from one of North America's areas most threatened by sea-level rise.

The Bayside World

Around the world's shorelines are numerous indentations, including bays, lagoons, and estuaries. Most of these bodies of water, especially on coastal-plain coasts, were formed by the flooding of the land by the last gasp of the last ice retreat that began 18,000 years ago. As the sea level rose as much as 400 feet, the land was flooded, and irregularities in the land surface such as valleys and ridges were inundated and became bodies of water separated by preexisting ridges. The two largest such bodies of water in the United States are the Chesapeake and Delaware bays, and in Australia, a bay similar in size and shape to the Chesapeake is Spencer Gulf.

The current sea-level rise in much of Chesapeake Bay is close to 1.75 inches per decade, among the highest rates on the U.S. east coast. The cause of this high rate of sea-level rise, is, in part, tied to a meteor that hit about 35 million years ago in what is now lower Chesapeake Bay. The nature of the mud and sand sediment that filled in the massive crater causes the land to compact and sink and the sea level to rise accordingly. Also contributing to a relatively high rise in sea level is the collapse of the bulge of land forced upward in front of the now disappeared glaciers from the last ice advance. It is estimated that during the twentieth century, Chesapeake Bay deepened by 1 foot.

Crisfield, Maryland, a colorful bayside village on the Chesapeake, once had a thriving seafood industry and called itself the seafood capital of the world. With the collapse of the fishing industry in the bay, the town of 2,700 remade itself as a small but successful tourist/retirement spot. It is clear that in low-lying Crisfield and many other bayside communities, the 1-foot increase in water depth has provided the basis for a very significant increase in storm-surge elevation. With or without future storm surges, the city will be "swimming with the fishes" if the anticipated 3-foot sea-level rise by 2100 materializes.

Already, the evidence of a rising sea is all around the town: advancing freshwater marshes occupying what were once lawns and farmers' fields, as well as the increasing extent of high tides, especially spring tide flooding.

Adding insult to injury, the offshore islands in the bay that once protected the town by reducing the size of the waves in storms, have shrunk in size and elevation or even disappeared. They are no longer there to trip up storm waves, so now the bay has a 30-mile distance (fetch) over which winds can build up the storm waves before they strike the Crisfield shore.

Hurricane Sandy in 2012 brought a great deal of flooding to Crisfield, and although no one died because community members rescued one another, the storm waters brought much soul-searching to the community afterward. David Montgomery reported in his article on Crisfield's response to Sandy that some people thought that cutting CO_2 emissions might save the community—a forlorn hope, since much of the climate change and related sea-level rise is already "locked in" and spells doom for this community. A local resident told Montgomery, with regard to scientific projections of the rising sea level, that it's one thing to sit in a lab and draw a waterline on a map, but what if you love that wet ground?

The moment for Crisfield residents to retreat passed. Federal money arrived. The Army Corps of Engineers proposed building a breakwater offshore that might reduce the waves but would do little to alter future flood levels. Nationwide, church volunteers from Lutheran, Baptist, Episcopal, and Mennonite disaster-response teams converged on the town. A few buildings were moved to higher ground, many were raised, and some were abandoned. But on the ground, nothing has changed fundamentally. Crisfield remains in grave danger. Furthermore, these actions likely place residents at risk for the guaranteed disaster down the road (as the waters continue to rise) by delaying the inevitable move away from this doomed community.

Cemeteries on the islands of Chesapeake Bay and its low-lying mainland shores, which originally were on dry land, are now submerged in marshes or completely gone, lost to both inundation and erosion by the advancing sea. In his book *The Disappearing Islands of the Chesapeake*, D. B. Cronin noted that more than 500 islands in Chesapeake Bay, mostly tiny and uninhabited, have disappeared since the arrival of the European settlers. One of those islands was Holland Island, just offshore from Crisfield. The island was once home to more than 360 people and dozens of fishing vessels. But the severe shoreline erosion that began in the late nineteenth century led to its complete abandonment by 1918. The last house toppled into the bay in October 2010.

Barrier Islands: Ribbons of Sand

Barrier islands are found primarily on low-lying, flat coastal zones, with wide, adjoining continental shelves and sediment dominated by sand. With few exceptions, the world's coastal plains, including in Siberia, Alaska, the United States' southeast and Gulf coasts, western Australia, the Pacific coast of Colombia, and parts of Mexico and Mozambique are lined by barrier islands.

There are 2,200 barrier islands around the globe, on every continent except Antarctica (table 8). They range from icy Arctic conditions (Shishmaref, Alaska, is on a barrier island), to the steamy hot equatorial coasts of Colombia, Brazil, and Nigeria. In all, barrier islands make up about 10 percent of the total length of the world's open-ocean shorelines.

These long, narrow islands, composed of sand or gravel, are capable of moving up and landward in response to the rising sea level, by means of a process called *island migration*. The islands

TABLE 8
Global Distribution of Barrier Islands in Four Oceans

Ocean Basin	Number of Islands	Total Length (miles)
Atlantic	506	5,163
Pacific	448	2,857
Indian	605	2,676
Arctic	475	1,635

range in length from a few hundred yards to more than 100 miles. Padre Island, Texas, at 113 miles, is the world's longest barrier island. Some, like the barrier islands on the Niger delta, are 3 miles wide, but most barrier islands are a half mile or less in width, and a few are less than 50 yards wide. Most of the islands belong to the United States (table 9), which is why most of the scientific and planning literature about the islands has been written by American researchers.

Just about everywhere in the world where barrier islands exist, they are much sought after for development. The major exceptions are the islands in the cold Arctic Ocean, where only a few small villages exist. For the most part, these islands are occupied only by summer hunting and fishing parties and are wild and free to migrate with sea-level rise.

In the warmer, richer parts of the world, barrier islands provide a beautiful view of the sea, a sea breeze, and a fine beach to stroll along, lie on, fish from, or just hang out on. In remote tropical areas such as the Pacific side of Colombia and the Atlantic coast of Nigeria, small villages often are located next to the sea—not for recreation but to take advantage of the breeze to reduce numbers of malaria-carrying mosquitoes, and to be near fishing grounds.

TABLE 9
Global Distribution of Barrier Islands by Country

Country	Number of Islands	Total Length (miles)
United States	405	3,054
Mexico	104	1,392
Russia	226	1,020
Australia	208	905
Mozambique	115	563
Brazil	72	559
India	45	476
Madagascar	119	377
Colombia	63	364
Nigeria	23	337
Vietnam	29	330
Canada	154	294

Barrier islands are among the most dynamic large land features, as they frequently are affected by storms and, in fact, require storms for their survival. Recently the Army Corps of Engineers declared that Fire Island, New York, had been "damaged" by Hurricane Sandy, a clear misunderstanding of barrier-island evolution processes. In fact, storms bring sand to the islands, raise their elevations, and widen their back sides, all in the process of island migration. This natural change is not damage. Instead, the islands' survival depends on these processes, even though they often create problems for people who choose to live on top of these dynamic systems.

Recently the Army Corps of Engineers declared that Fire Island, New York, had been "damaged" by Hurricane Sandy, a clear misunderstanding of barrier-island evolution processes.

Some islands are narrow and low, such as most of those on the Outer Banks of North Carolina, Pakistan, the Mississippi and Mekong deltas, and the Arctic islands of Siberia, the United States, and Canada. Often a reach of barrier-island coast is a mix of both narrow and wide islands. The narrow islands are more likely to be overwashed in storms and often are in the actual process of island migration. Not so for the wide islands, many of which are thinning by erosion on both sides. The human retreat already has begun on the Outer Banks of North Carolina (although it is not yet called "retreat"), as dozens of small buildings have fallen in during the last two decades and have not been replaced.

Buildings on wide and high islands such as the Sea Islands of Georgia and some of the Frisian islands of the Netherlands and Germany are likely to survive longer without relocation than are those on the narrow islands. But beachfront buildings, whether on low and narrow or high and wide barriers, are at very high risk.

As sea-level rise threatens more communities, even replenished beaches will eventually become a memory, except for those in a few wealthy island communities that can afford them. But with time, these replenished beaches will rapidly become unstable, and their lifespan will become shorter. This is because the *equilibrium shoreline* is somewhere behind the held-in-place beach that is now not located where nature wants it to be. Simultaneously, as more resources are required to maintain a beach, cities will have to absorb much of the funding for the response to sea-level rise. It is likely that cities will trump barrier-island communities for federal dollars.

Rocky Coasts

Rocky coasts, unlike barrier islands, are usually characterized by steeper coastal zones, with hills and mountains lining the shore on one side and narrow, steep continental shelves on the other. Broadly speaking, the west coast of North and South America is lined mostly by rocky coasts, whereas the east coast of these continents is dominated by sandy coasts and barrier-island chains.

The shoreline erosion problem is far more complex on rocky coasts than on those coasts with long, sandy beaches. The type of rock, which controls the degree of resistance to erosional forces, sometimes varies widely over short distances. Hence the response to storm waves can also vary widely, and of course, the response will vary with the direction, duration, and magnitude of storms. Besides the rock type, the presence of fractures, joints, and faults within the rocks is important, as they can determine erosion rates.

In an excellent article entitled "Lost Neighborhoods of the California Coast," Gary Griggs of the University of California at Santa Cruz outlines the rocky-coast problem of California. Typical of rocky coasts, this 1,100-mile-long coast includes lowlands associated with estuaries, lagoons, dunes and beaches, steep mountains, and various coastal cliffs and bluffs.

The important point that Griggs makes is that there are lost neighborhoods along the California shoreline just as there are along the east coast barrier-island shoreline, which receives the most media attention. The reason is that a single large storm along a sandy, low, flat coast, like that found on the east coast, can cause damage over distances of hundreds of miles. For example, the 1962 Ash Wednesday nor'easter caused damage from Massachusetts to northern Florida. In contrast, damage along a rocky coast would have been confined to low areas and erodible bluffs.

Griggs recognizes several different types of fluctuation in the shoreline position, including

Seasonal fluctuations in beach width. These seasonal changes are important along bluffed and cliffed coasts because the width of the beach at the toe of the bluff may determine how much storm-wave energy can be spent on erosion that causes bluff recession.

Long-term beach migration. In California, this landward shift of the beach position is commonly caused by a reduction of the sand supply to the beach, caused either by nearby engineering structures that trap sand (such as the aforementioned Santa Barbara breakwater) or by the loss of potential beach sand trapped behind dams. Loss of sand due to dams is a major cause of beach retreat on most rocky coasts, but on coastal-plain coasts, river sand is deposited at the head of estuaries and doesn't make it all the way to the beaches in large quantities. Thus, dams constructed upstream on rivers flowing to most of the U.S. east coast do not cut off significant sand supply to beaches, but not so on the west coast.

Even though loss of sand may be the underlying cause of long-term landward beach migration, the actual events that result in visible erosion, overwash, and beach migration are almost always storms. Newport Beach, California, is a sandy spit that behaves like a barrier island. The spit suffered severe erosion, especially in big storms in 1934, 1965, and 1968. The 1965 storms caused the beach to narrow by more than 160 feet. As on the barrier islands along the east coast, the response to the storm has been both the replenishment of beaches and the construction of hard structures like groins in an attempt to hold the beach in place.

Shoreline migration. This response of shorelines to sea-level rise causes steady landward movement. The measure of the shoreline position is the wet–dry line on the beach, and the rate of landward movement is controlled mostly by the slope.

Cliff and bluff erosion. The recession of cliffs (solid rock) and especially bluffs (unconsolidated sediment) is an irreversible landward shift of the shoreline. Globally, bluff recession is caused by a variety of factors (see chapter 11). Annual recession rates may range from a small fraction of an inch to several feet each year, and in the case of bluff collapse, hundreds of feet may retreat in an instant. Along the California coast, moving buildings back from the edges of bluffs and cliffs is a fairly common response. In 1983, a three-story building was moved back from a cliff top in Pacifica after the loss of 30 to 50 feet of bluff.

The rock type is clearly a factor in cliff retreat. At Gleason Beach, California, along a cliff top where rock types vary, houses have fared much better on resistant rock types, as opposed to those constructed on weak, highly fractured rocks. The presence or absence of a beach at the base is also a clear factor in erosional recession. A wide beach absorbs some of the wave energy that otherwise might strike the cliff. Other erosion factors include rainfall, groundwater seepage aided by water from septic tanks, and lawn watering, which lubricates slip planes (e.g., clay layers) that generate landslides.

Some bluff erosion is catastrophic. Griggs cited the loss, during an El Niño storm in 1983, of an entire row of mobile homes situated at the bluff edge in Pacifica, California. *El Niño* is a weather phenomenon that occurs every few years and affects the Pacific shores, bringing more severe storms and raised sea levels. The El Niños of 1982/1983 and 1997/1998 were particularly destructive.

The Pacific Institute published in 2009 a detailed summary of California's vulnerability to an expected 100-year storm after a sea-level rise of 4.6 feet. Included are 1,750 miles of roads and highways, 110 miles of railroad tracks, 332 regulated waste sites, 30 power plants, 140 schools, 94 health care facilities, 17 fire stations, and 17 police stations. Much of the likely loss would be from various kinds of mass wasting (i.e., slumps, landslides, rock falls, and debris flows).

In contrast to California, the bluffs of Puget Sound, Washington, have a somewhat different geologic and oceanographic setting, but they also frequently fail, causing considerable property loss and the disruption of transportation systems. The Puget Sound bluffs, north of Seattle, often fail because of a clay layer at their base. Rainfall saturates the overlying sediment, adding mass, but also acting as a lubricant on the clay, allowing the bluff, sometimes with houses on it, to flow toward the sea by slumps and slides.

Ironically, this retreat of bluffs in Puget Sound is an important source of sand and gravel for beaches. Blocking off the base of a bluff with a seawall, therefore, starves beaches of their sediment and may actually accelerate the bluff's retreat because of the protection afforded by a beach.

On shorelines in northern latitudes, including Scandinavia, New England and eastern Canada, and on beaches from Puget Sound north, unconsolidated sands and gravels left behind by the glaciers provide much material for beach construction but are particularly vulnerable to erosion and failure by sliding.

As this quick survey of coastal environments has shown, development on all coasts is at risk in a rising sea. All indications are that inundation and storm surges will be ever more severe and costly in the future, amplified by increasing coastal populations and a rising sea level.

8

Drowning in Place

Infrastructure and Landmarks in the Age of Sea-Level Rise

A society runs on its infrastructure, the basic physical and organizational structures and facilities needed for human settlement and security. Inundation from sea-level rise, storm surges, erosion, flooding, and severe storms is affecting the places where we collect and treat waste, generate power, and store oil supplies and where we locate our military bases, ports and cities, highways, rail corridors, and the places where our cherished historic landmarks are situated.

Sea-Level Rise and Sewage

Hurricane Sandy developed as a fierce, devastating confluence of events that battered the east coast of the United States in 2012. As Sandy began arriving in the mid-Atlantic region, it became wedged between a stationary cold front over the Appalachians and a static high-pressure air mass over Canada. Together, these systems blocked Sandy from moving in a more typical storm fashion either north or east. Sandy intensified and came ashore northeast of Atlantic City, New Jersey, and continued northward into New York.

The potential for damage in New York City is huge. This is because the city has 520 miles of varied coastline—beaches, marshes, inlets, bays, barrier islands, estuaries, and rivers—that is longer than the coastlines of Miami, Boston, Los Angeles, and San Francisco combined.

Hurricane Sandy flooded 51 square miles of the city (17 percent of the city's total land area) and inundated some 88,700 buildings, structures that included more than 300,000 residences. More than 433,000 New Yorkers live in the areas flooded by Sandy. Subways, trains, and air transportation were paralyzed as Sandy made landfall. Hospitals flooded; power and water to tens of millions of people were shut down. One foul impact of Hurricane Sandy came from the inundation of sewage-treatment-plant infrastructures in at least nine different states. The storm surge swamped and shorted out electrical equipment, forcing sewage to back up in the system and overflow in unintended ways. Other treatment facilities were overwhelmed by the sheer volume of water in their systems. Even though plant operators rallied to try and control the systems, some were forced to send an estimated 11 billion gallons of untreated or partially treated sewage into rivers, bays, canals, and city streets. Household waste from toilets, kitchens, and showers; storm-water runoff from streets; and oil and grease waste from industrial sources all were mixed into a "witches' brew."

Unfortunately, Sandy is not an isolated incident—any large storm can create similar infrastructure problems. The United States has an aging sewage and wastewater infrastructure, built over time to reduce water pollution and to protect the health of a much smaller population. Sewage-treatment plants are increasingly vulnerable to failures in the era of sea-level rise. When exposed to storm surges, waterlogged wastewater-treatment plants can remain incapacitated long after the rain or floodwaters recede.

Treatment plants are usually built in low-lying areas close to water, so that raw sewage can be piped to the plant for treatment. The low-lying placement lets gravity aid in this work, after which the treated sewage is sent

Treatment plants are usually built in low-lying areas close to water, so that raw sewage can be piped to the plant for treatment. The low-lying placement lets gravity aid in this work, after which the treated sewage is sent to the receiving waters (streams, rivers, bays, sounds).

to the receiving waters (streams, rivers, bays, sounds). But the same receiving waters also happen to be our places for recreation, so they expose populations to pollution and disease if the system doesn't work properly.

Unlike housing and transportation infrastructure, which, in a retreat, are technically movable, the very function of sewage plants all but requires them to be located near waterways. As the seas rise, the resulting increase in these types of pollutants can worsen already unhealthy beaches or degrade more pristine areas, which adds to the impetus to retreat from the shore. Obviously the potential for sea-level rise should immediately be included in the planning and permitting for these treatment systems. Yet despite this being painfully obvious, cities like Miami may not plan accordingly. In his June 2013 *Rolling Stone* article "Goodbye, Miami," Jeff Goodell wrote that Miami has no plans to relocate its decrepit wastewater-treatment plant at Virginia Key on Biscayne Bay—a plant that has been plagued by spills over the past decade. Instead, Miami is planning to sink $550 million in repairs and upgrades. The plan is to prepare the plant for a 3-foot rise in sea level, but the structure remains a vulnerable sitting duck on the low-lying barrier island.

Energy Infrastructure

In 2012, the nonprofit research group Climate Central mapped the location of energy facilities in the continental United States situated at elevations 4 feet or less above the high tide line. The group identified nearly 300 facilities, including 130 natural gas, 96 electric, 56 oil, and 4 nuclear power facilities. Louisiana, in particular, has an abundance of natural-gas facilities at low elevations. The study didn't look at whether these facilities had intervention measures such as seawalls or levees that could offer some temporary protection against the rising seas. Nonetheless, it is clear that the permit-siting decisions we are making today will exacerbate the problem in the future.

In Florida, both the Turkey Point Nuclear Generating Station and the St. Lucie Nuclear Power Plant on Hutchinson Island are located on barrier islands. Turkey Point is a twin-reactor nuclear power station east of Homestead, Florida, which was directly hit by Hurricane Andrew in 1992 and had to be shut down. The plant was built to withstand winds up to 235 miles per hour, a standard that exceeds the maximum winds historically recorded by Category 5 hurricanes. Even so, the storm damaged the plant's main water tank, severely damaging the water-treatment plant and a smokestack of one of the site's buildings. The smokestack had to be demolished and was rebuilt. This is an example of the head-in-the-sand mentality of some South Floridians, as shown by the response of Turkey Point's lawyers to expressions of concern about the power plant's low elevation. They argued that any decisions concerning responses to sea-level rise should come from the state capital in Tallahassee. Ironically, the Florida legislature has been accused of responding to the global climate-change issue with the attitude that if things get bad, the federal government will step in. The buck is passed around and could continue to be passed until it just floats away.

Hurricane Sandy caused the shutdown of several coastal nuclear power plants, including the Jersey Shore's Oyster Creek Nuclear Generating Station, which took the brunt of the storm surge. Much like sewage-treatment plants, nuclear power plants tend to be built near low-lying coastlines in order to have access to sources of water to cool the reactors. These plants also rely on the electric grid to power their cooling systems. During a storm, this presents a double whammy: exposure to flooding, plus power outages that endanger the safe operation of the plant.

Much like sewage-treatment plants, nuclear power plants tend to be built near low-lying coastlines in order to have access to sources of water to cool the reactors. These plants also rely on the electric grid to power their cooling systems. During a storm, this presents a double whammy: exposure to flooding, plus power outages that endanger the safe operation of the plant.

Sandy came after the meltdown in 2011 of Japan's Fukushima Daiichi Nuclear Power Plant, which the world watched in horror and disbelief. The Fukushima reactors lost power from the grid after tsunami waves came ashore, and the backup diesel generators failed because of flooding. This dreadful event should have accelerated efforts worldwide to manage nuclear plants for flooding events. After Fukushima, the U.S. Nuclear Regulatory Commission ordered nuclear facilities in the United States to review and update their plans for seismic events and potential flooding from sea-level rise and storm surges.

Radioactive spent nuclear fuel is stored on-site at nuclear power plants across the United States, which is meant to be a temporary solution until a permanent nuclear-waste repository is built. *Waste confidence* refers to the Nuclear Regulatory Commission's confidence that nuclear waste can be permanently disposed of when necessary. Although Congress directed the U.S. Department of Energy in 1982 to construct a deep geologic repository for high-level radioactive waste at Yucca Mountain in Nevada, that state's legislators opposed the repository, and the department withdrew its license application in 2010. In the absence of a permanent repository, waste will be stored on-site indefinitely at numerous facilities and remain dangerous for thousands of years. A permanent site such as Yucca Mountain would remove the waste far away from the rising sea and reduce at least one of the threats to energy infrastructure in the United States.

Oil Refineries

According to NOAA's State of the Coast Report in 2013, more than one-half of the United States' energy is produced in 30 coastal and Great Lakes states. More than one-quarter of U.S. crude oil is extracted from state and federal offshore waters, although this is rapidly changing with the production of shale oil fields in the inland areas. Roughly two-thirds of all U.S. oil imports pass through the Gulf Coast, which is home to seven of the country's 10 largest commercial ports. In 2005, Hurricanes Katrina and Rita destroyed roughly 100 offshore oil-drilling platforms and damaged nearly 600 oil and natural-gas pipelines.

Although oil platforms usually can be jacked up in response to a rise in sea level, oil refineries cannot. In 2014, the U.S. Energy Information Administration estimated that the United States has 142 operable refineries, three of which are inactive. Storm surges buoyed by sea-level rise near refineries cause oil spills, damage the large-capacity storage tanks holding millions of gallons of oil, and threaten our nation's oil supplies. It is ironic that the very refineries producing the gas and oil that release carbon dioxide and methane that raise the sea level are endangered by sea-level rise.

The shoreline of the Delaware City Refining Company along the Delaware River is disappearing in front of the refinery. The company filed an application in May 2014 under the Coastal Zone Management Act seeking to armor against the rising sea levels. On its application, the company wrote, "The extent of the shoreline erosion has reached a point where facility infrastructure is at risk. . . . The extent of tidal encroachment [meaning sea-level rise] is obvious, . . . [and] A review of historical aerial photography suggests that the rate of shoreline erosion is increasing."

The refinery then asserted that the only solution was to build natural sand dunes and a protective ring of buoys with the "resilience to deal with sea-level rise for at least 50 years" (a doubtful assertion). The plan would sink dozens of pilings and hundreds of large, four-sided, wave-calming blocks offshore of the refinery. One problem with this plan is that the construction of the project could direct storm surges toward Delaware City, a small coastal community nearby. Indeed, the Sierra Club commented as such to the permitting authority in 2014: "Severe storms are eroding the shoreline and affecting the business of an oil refinery, Delaware City Refining Company, that is threatened by increasing extreme weather. In other words, climate disruption is hitting the doorstep of its source (CO_2) and it is seeking taxpayer-funded shoreline protections due to tidal encroachment—which is one way of saying sea-level rise."

The refinery is owned by PBF Energy and is one of the state's largest employers, but the company would be using public money to benefit a private enterprise. Perhaps more should be made of the fact that an oil company is now forced to confront the impact of sea-level rise and climate change. In fact, the fossil-fuel industry is

In fact, the fossil-fuel industry is actively planning for climate change, on the one hand, while actively denying its existence or downplaying its potential impact, on the other hand.

actively planning for climate change, on the one hand, while actively denying its existence or downplaying its potential impact, on the other hand.

Transportation

Road, rail, and airport delays and travel closures are increasing, and evacuation routes for coastal areas are washing out. NOAA reminds us that the potential exposure of the transportation infrastructure to flooding is immense (from its website "State of the Coast"). Along the northern Gulf Coast, an estimated 2,400 miles of major roadway and 246 miles of freight rail lines are at risk of inundation by a sea-level rise of 4 feet.

Several of the largest airports in the United States have at least one runway with an elevation within 12 feet of current sea levels. These include LaGuardia and JFK in New York and the airports in Philadelphia, Washington, D.C., San Francisco, and New Orleans. Often, extensive oil tank-farms are situated at these airports to supply fuel for the planes, and some airports have already begun to assess their vulnerabilities and decide on action.

The U.S. Military

Scientists and the U.S. military have argued for years that climate changes will spur waves of human migration and battles over increasingly scarce resources, particularly water. Every four years the U.S. Department of Defense publishes its *Quadrennial Defense Review* (*QDR*), focusing on security threats around the world. In 2010, the *QDR* named climate change "an accelerant of instability or conflict." The department determined that the U.S. military would be increasingly pressured to support civil authorities responding to extreme weather events and that climate change is negatively affecting the military's access to land, air, and sea training and the test space on which the military relies.

More than 30 of our military installations are considered to be at risk from rising sea levels, though they were not named in the review. One of these is undoubtedly Norfolk, Virginia, where the U.S. Navy has had to add two levels to its docks because of sea-level rise. In addition, the military is very much aware that the environmental refugee problem caused by the rise in sea level and that the movement of millions of people, sometimes across national boundaries, may involve military action or participation. Accordingly, the military and national security agencies appear to be at the forefront in the United States in recognizing and planning for the inevitable sea-level rise.

North Carolina is known for its numerous coastal military installations, including Camp Lejeune and the Dare County Bombing Range. Home to multiple military commands, Camp Lejeune covers more than 146,000 acres, including 11 miles of beach to support the amphibious training of troops. Set aside for the Dare County Bombing Range, a military facility considered vital to our national security, are more than 46,000 acres of marshland, forests, and open space. Several projects and studies relating to sea-level rise have focused on these two and other military installations, and in its current planning phase, the military has chosen to take sea-level rise head-on, in considerable contrast to the state's civilian leadership.

The National Aeronautics and Space Administration and the Conflict of Wallops Island

By contrast, the National Aeronautics and Space Administration (NASA) improbably seems to be incapable of responding to the very data it was tasked to gather on the rising sea levels around the globe. This is described in fascinating detail in an installment in the 2014 Reuters series on sea-level rise.

Wallops Island, Virginia, is part of the beautiful winding chain of barrier islands on the eastern seaboard of the United States. A series of open bays and salt marsh, the southern extension of Chincoteague Bay separates Wallops Island from the mainland. NASA's billion-dollar test-flight facility is located on the southern end of the island and is used by a number of agencies, including the navy, NOAA, and commercial enterprises. More than 16,000 rockets have been fired from Wallops Island since World War II. A number of satellites with the specific mission of documenting the global sea-level rise and the melting of the world's ice sheets have also been launched from here. Nearby, planes from the International Arctic Research Center record the pace of melting of the world's ice sheets.

The shoreline at Wallops Island has undergone at least 150 years of documented chronic erosion. The first seawall was constructed in the 1940s and, after storms, has been extended, augmented, and repaired numerous times over the years. By the early 1970s, about 50 wooden groins had been built. Assawoman Inlet, a small inlet at the southern tip of the island, closed sometime in the late 1980s and no longer exists. In the 1990s, NASA built the current 14-foot-high rock seawall.

In January 2009, a federal interagency assessment of the mid-Atlantic coast found that Wallops Island (along with Assawoman, Metompkin, and Cedar Islands) likely had crossed a "geomorphic threshold" from a relatively stable state into a highly unstable condition, one in which rising seas could trigger "significant and irreversible changes." In geologic terms, the islands are beginning to migrate and will likely shrink in size and be broken up by inlets. In human terms, this island is in trouble. Ten months after the report, with its blunt and startling conclusions, construction started on a $15.5 million rocket-assembly building and a $100 million launch pad located a mere 250 feet from the pounding surf on the southern end of the island. The building phase was completed in 2012, two months before Sandy roared by.

In recent years, Wallops Island has been losing an average of 12 feet of shoreline a year. In early 2010, NASA proposed a $43 million project to extend the seawall by about 1,400 feet and to build a new 4-mile-long replenished-sand beach characterized as a "fifty-year design-life storm damage reduction project," an expected longevity that is far off the scale of reality in an age of sea-level rise. The seawall and beach-replenishment projects were approved, and work was finally finished in August 2012. Sandy damaged the new wall and washed away one-quarter of the 3.2 million cubic yards of new beach sand. An $11 million beach-replenishment project to replace the sand lost in the hurricane, paid out of the Sandy relief fund, began in July 2014.

As required by law, NASA had released a draft environmental impact statement (EIS) on the plan. The justification for the project was to reduce the need for the emergency repairs and the potential for storm-induced physical damage to the more than $1 billion in federal and state assets on Wallops Island.

The EIS reviewers criticized the project design's inadequate consideration of rising sea levels and assessment of their impacts, despite the passage recognizing the "geomorphic threshold." NASA responded by nearly doubling (in the impact statement) the number of references to the effects of sea-level rise. But in the official summary record of the decision, which announced that Wallops would proceed with the project, sea-level rise isn't mentioned anywhere. The head-scratching explanation, according to the Reuters articles on sea-level rise, is that NASA explained that it had whittled the issues "down to only the highest points," and sea-level rise wasn't among them.

As part of the same project, a 500-foot-long groin was proposed for the south tip of the island that would trap sand and widen the beach. The proposal was similar to the "spite groins" of the Jersey coast, where several towns placed their longest groins at the south or downdrift end of town to capture as much sand as possible before it could reach the next town downstream. A loud outcry from the environmental community and a complaint from the Nature Conservancy halted this Wallops Island groin project, as the groin was certain to further erode the conservancy's 14-island Virginia Coast Reserve to the south.

In addition, NASA is considering adding another launch pad, enforcing its attitude toward sea-level rise that is reflected by its decision making on Wallops Island. Such a decision by this, the very federal entity that is most aware of the future of the sea level, bodes poorly for the nation's future response to sea-level rise.

According to the Reuters report, just a few hundred yards away from the northern tip of Wallops Island is another example of people's reluctance to change their habits in response to sea-level rise. In this case, the government agency, unlike NASA, is trying to do the right thing but is frustrated by local people who fear the impact of change on their livelihood. The very popular beach at Chincoteague Wildlife Refuge on a narrow spit at the south end of Assateague Island, Virginia, has a shell/gravel parking lot that can handle 1,000 cars. The parking lot has been damaged repeatedly by a combination of storms and erosion at the rate of 10 to 20 feet per year. In order to avoid the cost of repeated repairs and to maximize protection of the natural environment, the U.S. Fish and Wildlife Service has proposed replacing the beach parking with a shuttle system starting in the nearby town of Chincoteague. The locals exploded at this idea, claiming it would destroy the local economy. After much bitter debate, the agency agreed to build a smaller and temporary parking lot on a more stable beach, but it did not promise to repair storm damage to the new lot. Even when no buildings are threatened, the pressures to hold the line against sea-level rise can be insurmountable.

Our Human History: Landmarks

Human culture is made up of beliefs, behaviors, places, and objects that are held in common. An astonishing number of the places that define the world's culture, and the cherished monuments we have created to celebrate that history, are under threat from sea-level rise. Jamestown, Virginia, the site of the first English colony in the United States, is prominent among them. Some landmarks and the perils they face are described in a report by the Union of Concerned Scientists entitled "National Landmarks at Risk: How Rising Seas, Floods, and Wildfires Are Threatening the United States' Most Cherished Historic Sites." It is certain in the long run that most of the precious sites near coastal waters will be lost in this or the next century as the sea level rises.

Fort Fisher, North Carolina, built in 1861, protected the channel in the Cape Fear River leading to Wilmington, and the fort enabled the Confederates to be resupplied until the Northerners captured it in 1865. What is left of it today is mainly earthworks that were constructed by the Confederates, along with a large stone and concrete seawall that now protects the fort. Special permission to construct the wall was granted because of the fort's historical significance.

But someday, decades from now, walls won't hold the shoreline in place, and the fort will be lost.

Beauvoir Estate (the home of Jefferson Davis, president of the Confederacy) in Biloxi, Mississippi, is a beachfront plantation that remained unscathed through a number of storms but was severely damaged in 2005 by Hurricane Katrina. It now has been restored but is very vulnerable to the future rise in sea level.

Another type of beachfront treasure is Turtle Mound, a prehistoric archaeological site of enormous cultural value, located south of New Smyrna Beach, Florida. Listed on the U.S. National Register of Historic Places in 1970, the area later became part of the Canaveral National Seashore, which is primarily a barrier-island ecosystem with water as a dominant feature. The turtle-shaped mound contains oysters, tools, bones, pottery, and other refuse left by the Timucuan people, in piles that became mounds called *middens*. Archaeologists believe that people may have also used this site as a high-ground refuge during hurricanes, and early sailors used Turtle Mound as a navigational device because it could be seen from far out at sea. The mounds were believed to have been as much as 75 feet tall, but today are only 40 feet tall, reduced through time by compaction and human disturbance.

In response to the rise of the sea and erosion, Canaveral National Seashore and the U.S. National Park Service have partnered with the University of Central Florida to test stabilization methods and to plant a "living shoreline" of submerged mats of oysters, spartina grass, and mangroves. The principle of living shorelines is restoration and protection through the use of natural materials, including marsh plantings, shrubs and trees, limited use of low-profile breakwaters/sills, strategically placed organic material, and other techniques. Living-shoreline techniques are sometimes considered as an alternative to hard armoring, but most such shorelines do contain a strong element of hard stabilization along the outer margin of the project. One problem is that the hard armoring at the margin of the living shoreline typically destroys the beach. Estuarine beaches are often small and covered with debris, yet they are an important part of the local ecosystem. Advocates of "living shorelines" need to define and regulate this term more precisely so that it is not misused simply to allow more unnecessary and damaging hard stabilization of estuarine shorelines.

Lighthouses

Perhaps the best-known lighthouse is the United Kingdom's Eddystone Lighthouse, built in 1696 and made famous in song and legend. Henry Winstanley, the designer, assured all concerned that it would withstand any storm, so during the Great Storm of 1703, he and, at his insistence, five lighthouse workers remained in the lighthouse. Unfortunately, the lighthouse was completely destroyed in the storm, and all six men were killed. Winstanley's lighthouse was the first of four at this location, the first three all falling victim to storms.

The Point Isabel Lighthouse, facing Laguna Madre, Texas, is the southernmost lighthouse along the state's shoreline. It is just one of hundreds of lighthouses along all the U.S. (and the world's) coasts. Many, if not most, are endangered by erosion and sea-level rise. Some of the lighthouses have been moved back from a retreating shoreline, and interestingly, the Hunting Island State Park Lighthouse in South Carolina was designed to be taken apart and moved. In 1889, the steel plates forming the wall of the lighthouse were disassembled and reconstructed at a new site 1.25 miles away from the original site.

The most famous example of such a move is the 3,500-ton Cape Hatteras Lighthouse, which was rolled back 1,500 feet from the eroding North Carolina shoreline in 1999, a move that then cost $12 million. The locals were strongly opposed to the move because they thought that it would damage the tourist industry, and many believed that the lighthouse could not be moved. Ironically, it is now more of a tourist attraction than ever.

The Morris Island Lighthouse in South Carolina was built on the back side of Morris Island, about 1,200 feet from the beach. The Charleston jetties constructed in the 1890s caused severe erosion on the island, causing it to migrate right out from under the lighthouse that now stands 1,200 feet offshore.

Sea-level rise could be mitigated locally in the short term by managing natural processes to build new soil elevation and encouraging plant and animal communities to adapt to sea-level rise by migrating. But that is just in the short term. Alternatively, we could continue to engineer methods to artificially and futilely attempt to hold back the sea. In the long term, we will have to retreat from the shoreline, and accept that the loss of important and cherished sites, including the Turtle Mound, is inevitable.

9

The Cruelest Wave

Climate Refugees

The flight of humanity from the shore is an inevitable consequence of sea-level rise. The world's waters are rising in a measurable, continuous process that will force large numbers of refugees to flee their communities for safer ground. The human flight from climate change is bleeding quietly now but verges on hemorrhaging, as it is with today's political refugees. Unfortunately, climate change and sea-level rise are too infrequently discussed in terms of justice or morality for vulnerable populations, changes that raise some profound ethical issues.

Migration is an age-old response to a change in human conditions. Remaining mobile has always been a way to survive. *Adaptation* refers to human mobility either inside national borders (displacement) or outside (migration), and either way, it is rarely an easy path.

In 2016, in accordance with a series of sweeping plans, Europe had to adapt to a refugee crisis that escalated as a wave of terrified humanity flowed across the European border, fleeing violence in Syria, Iraq, and other countries. At a time when politics in Europe is particularly hostile to migrants, the crisis began with a war targeted at civilians, which resulted in their horrifying journey by foot and by boat. Exploitative networks of criminals move refugees for a fee but offer little humanity, landing them into overcrowded migrant camps and often leading them to hastily built fences, border closures, and a world crisis. This has been the worst refugee emergency since World War II but similar in the sense that the crisis is global. According to filmmaker and environmental advocate James Cameron and others, at the heart of the Syrian crisis lies a climate-related issue. The chain of political and violent consequences in this part of the world began with social reforms coinciding with a drought and the subsequent movement of farmers into the city slums as farms collapsed. Corruption and a lack of action by the government led to a void that was filled with rebellion and violence. This crisis is an indication of what may come if we choose not to plan for the inevitable environmental retreat but wait too long and then respond. Elon Musk, the cofounder of the electric carmaker Tesla, argued in a speech delivered in Berlin in September 2015 that climate change will trigger a refugee crisis that will dwarf the current political-refugee crisis.

There is one important difference between political and environmental refugees. Political refugees tend to move as individuals or groups that share a fear of persecution from a shared ideology, religion, war, or another driver. In the future, entire villages, towns, or island nations could flee sea-level rise, regardless of their differences or likenesses. Helping these affected people first will require a clearer definition of what exactly an environmental refugee is. Are any nations ready or able to take in entire populations without separating them, so that a welcomed refugee population could retain its cultural and linguistic integrity?

Political refugees tend to move as individuals or groups that share a fear of persecution from a shared ideology, religion, war, or another driver. In the future, entire villages, towns, or island nations could flee sea-level rise, regardless of their differences or likenesses.

Ideology, religious intolerance, and economic fear are the primary causes of resisting the acceptance of immigrants. For example, in the 1930s, the Midwest underwent a period of severe, prolonged drought, later known as the Dust Bowl. At the same time, the Great Depression, the drought, unusually high temperatures, poor agricultural practices, and the resulting wind erosion all contributed to the environmental crisis. By 1932, 14 *black blizzards*, or severe dust storms, had been recorded. A year later, this number increased to nearly 40. In his book *The Worst Hard Time*, Timothy Egan wrote, "In those cedar posts and collapsed homes is the story of this place: how the greatest grassland in the world was turned inside out, how the crust blew away, raged up in the sky and showered down a suffocating blackness off and on for most of a decade." This crisis set people in motion.

By 1940, 2.5 million people had fled the Plains states, most relocating to California. Even though this displacement was inside a nation rather than across borders, the migrants (known derogatorily as "Okies") were not welcomed. More workers moved west than the number of available jobs, tensions grew, and public-health concerns rose as California's infrastructure became overtaxed.

When the flow of humans is across national borders, the fear and resistance are amplified. Whereas a short-term sea-level event like a storm surge leads to only a temporary relocation, the world faces something very different with the long-term sea-level change. In richer communities, the rising seas challenge permitting and development decisions that have long provided a coveted coastal lifestyle to the millions who built near the ocean. Conversely, in poorer communities around the world, sea-level rise is a multiplier for the existing underlying stressors and vulnerabilities of poverty, isolation, and/or overpopulation. Well before complete inundation, poorer communities find that the intrusion of saltwater has contaminated their drinking water and made sustainable agriculture impossible. The rising seas in these communities threaten the ability of whole populations to remain in their homes, to continue in their traditional livelihoods and sustain their culture, and to count on the availability of safe food and water. Both health and prosperity will be increasingly at risk, and some nations will likely become uninhabitable even before they become inundated. The predicted flow of rising-sea refugees will come from coastal cities and low-lying islands, river deltas, and the African and Arctic regions alike. Although this flood of refugees is inevitable, we can choose whether this movement is planned or chaotic.

The Alliance of Small Island States (AOSIS) is an intergovernmental group of about 40 low-lying countries highly vulnerable to the impacts of climate change. Its slogan "1.5 to Stay Alive" refers to its goal of limiting the increase in global temperature to less than 1.5 degrees Celsius (3.6 degrees Fahrenheit). The AOSIS members are united in that they stand to lose their lands as the sea level rises. Among the most immediately threatened island nations are Kiribati, Tuvalu, and the Marshall Islands in the Pacific Ocean and the Maldives in the Indian Ocean. Their residents already are making plans for their adaptation or migration, negotiating with other countries, and making land purchases. In 1990, the Intergovernmental Panel on Climate Change (IPCC; see chapter 7) recognized this impact of climate change, and its March 2014 report states that "climate change can indirectly increase risks of violent conflicts in the form of civil war and inter-group violence by amplifying the underlying stressors." Predictions have varied from 50 million to 200 million environmental refugees by 2050, predominantly from the least developed countries. Even so, climate-change deniers pounce on the numbers in order to delay and divert the conversation. Although caution should be used when discussing the numbers, the reality is that the world is facing an immigration crisis.

Among the most immediately threatened island nations are Kiribati, Tuvalu, and the Marshall Islands in the Pacific Ocean and the Maldives in the Indian Ocean. Their residents already are making plans for their adaptation or migration, negotiating with other countries, and making land purchases.

Under international agreements, environmental or climate refugees are not recognized as having a legal status, and so they do not receive the same organized assistance that political refugees are afforded. An international convention in 1951 defined *refugees* as persons with a well-founded fear of persecution due to their race, religion, nationality, political convictions, or social class. According to the treaty, courts and administrative tribunals may grant rights in a new country to refugees meeting these criteria. The United Nations has recently begun defining a new status for environmental refugees. To that end, a researcher at the United Nations, Essam El-Hinnawi, popularized the term *environmental refugees* to describe people forced to leave their traditional habitat because of a marked environmental disruption that jeopardized their existence or quality of life.

In *A Perfect Moral Storm: The Ethical Tragedy of Climate Change*, Stephen Gardiner writes about the moral issues raised by climate change. He observes, for instance, that the world's most affluent nations may be tempted to pass on the cost of climate change to the poorer nations, so that the nations contributing the least to the human causes of sea-level rise would be hurt the most. Although this may not be done consciously, Gardiner argues that it may happen through a lack of action or will, because of the public's poor grasp of science and as a result of the complex human relation to nature. James Cameron, conversely, has argued that the shift against the deniers we are seeing now in the United States is coming from the people up, forcing our leaders to act.

The Small Islands

Ioane Teitiota, a Pacific islander from Kiribati, a nation of more than 30 islands, is sometimes called the world's first prominent environmental refugee. It is certainly true that Teitiota brought some of these moral and ethical issues to court and to the media. He migrated to New Zealand in 2007, contending that life in Kiribati with his wife was no longer possible because of the rising seas and threatened water supplies.

Teitiota and his wife filed for refugee status in New Zealand after overstaying their visas. First appealing to New Zealand's Immigration and Protection Tribunal, Teitiota argued that beginning around 1998, king tides regularly breached the seawalls around his village, which was overcrowded, and had fouled the drinking water. His village had no sewage system. In a historic and pioneering bid for asylum, Teitiota maintained that returning to Kiribati would endanger the lives of his children.

Teitiota's argument was that his flight was an indirect form of persecution because climate change is caused by humans. After years of court arguments, his bid for asylum was finally rejected in 2014. John Priestley, the high court justice presiding over Teitiota's case, called it "novel" but ultimately "unconvincing," explaining that a bid for asylum based on global-warming concerns did not meet the terms of the 1951 United Nations Convention Relating to the Status of Refugees. "By returning to Kiribati, he would not suffer a sustained and systemic violation of his basic human rights such as the right to life . . . or the right to adequate food, clothing and housing," Priestley wrote. He also postulated that "the economic environment of Kiribati might certainly not be as attractive to the applicant and his fellow nationals as the economic environment and prospects of Australia and New Zealand. But he would not, if he returns, be subjected to individual persecution."

Teitiota and his wife have run out of appeals. Their three New Zealand–born children also will be deported because New Zealand doesn't recognize the children of illegal immigrants.

Jeffrey Goldberg, writing for *Bloomberg Businessweek*, described life in the Kiribati islands as being in the shadow of a much-feared but likely necessary relocation of an entire society. None of the islands rises more than a few feet above sea level. The 103,000 people of Kiribati live near exposed shorelines, including 51,000 in the capital city of South Tarawa sitting on top of a very shallow lens of freshwater for drinking.

In the Tarawa village of Te Bikenikoora, the community meeting house was recently flooded in a king tide with no concurrent storm, but fish were deposited inside the building. Goldberg remarked on the tightness of the community, an essentially classless society distressed at the thought of being separated by relocation. Accumulating wealth is typically not important there; one employed person in a family group supports the others. But support for all the islanders is on the way. In an attempt to resolve the problems brought about by sea-level rise, the government of Kiribati has purchased 6,000 acres in Fiji, a volcanic island with high elevations, which may be used to supply food for the islanders or may eventually become a relocation site.

Perhaps the first atoll people to evacuate as a group are those of the Carteret Islands. Their evacuation began in 2003 and continues to this day. These islands are flooding at a rapid rate because the sea level is rising and the land is tectonically sinking at the same time. The relocation is proceeding slowly because of the inhabitants' reluctance to move from an atoll to a jungle.

AOSIS members Tuvalu and the Maldives also are in short-term danger. Tuvalu is made up of eight tiny coral atolls, with a total land area of just 10 square miles. The highest point in Tuvalu is 14.7 feet above sea level. The Maldives has a total population of 394,000 on around 1,200 islands, with 80 percent of the islands having elevations of less than 3 feet above sea level.

New Zealand has come to an agreement with Tuvalu to accept its 11,600 citizens in the event that rising sea levels overtake the country. The relocation will take place over decades. Meanwhile, leaders in the Maldives are working with other leaders in Australia, India, and Sri Lanka to plan an evacuation program to be carried out when the Maldives becomes uninhabitable. The former president of the Maldives was actively pursuing the possibility of buying land to which his people could migrate, but the current administration seems a bit less concerned about sea-level rise.

Bangladesh

Almost 1,800 miles north of the Maldives, Bangladesh is a poor, low-lying nation being inundated along its 440 miles of coastline, posing both humanitarian and political threats as the 14 million people who live near the current sea level seek higher ground. Even though Bangladesh's contribution to climate change is deemed negligible, the country is one of its

Even though Bangladesh's contribution to climate change is deemed negligible, the country is one of its principal victims. Three majestic Himalayan rivers converge in Bangladesh and meander to the sea through the Ganges Delta—beautiful to behold but less than ideal for flood control.

principal victims. Three majestic Himalayan rivers converge in Bangladesh and meander to the sea through the Ganges Delta—beautiful to behold but less than ideal for flood control. The country is a massive floodplain, with more than 20 percent of its land awash every year. As one of the world's least developed countries, it can ill afford the technology that others use to mitigate the effects of flooding and thus has to turn to less expensive means, such as creating houses built on stilts in coastal areas.

In a report by the CENTRA Technology and Scitor Corporation (quoted in Mckenzie Funk's book *Windfall*),

> Anticipated inundation and salt water intrusion in the Ganges Delta may displace tens of millions more Bangladeshi immigrants. India would not have the resources to cope with Bangladeshi immigrants pushing into West Bengal, Orissa, and the Northeast. . . . About half of Bangladesh's population, unable to sustain themselves through agriculture, will migrate to cities by 2050, and most of the migration will probably be to India. In addition, major disruptive events such as cyclones may generate mass refugee movements into India on much shorter timescales than the overall shifts in climate.

The migrations will be spurred on by the impact of storms such as the cyclone of 1970, which is estimated to have killed up to 500,000 people. Eight of the top 10 deadliest tropical cyclones in history have been in the Bay of Bengal, six of which hit Bangladesh.

India and Bangladesh share a border more than 2,500 miles long. Pakistan was partitioned by the British in 1947, and its border with Bangladesh (formerly East Pakistan) has little geographical or ethnic logic. Twenty-eight Bangladesh districts share a border with the Indian states of West Bengal, Assam, Meghalaya, Mizoram, and Tripura. Indeed, Pakistan's first prime minister called the new Bangladesh "moth-eaten," as it is separated into two regions by India. Even though the people on both sides share cultural, linguistic, religious, economic, and family ties, a tall, meandering, and partially reinforced border fence has been under construction in India since the 1980s and is guarded by soldiers. Some have dubbed it the Great Wall of India.

If it were fully completed, this border fence would be 2,100 miles long. By comparison, the Great Wall of China, which is no longer a border wall, is 3,889 miles long. Part of the challenge of building a fence between India and Bangladesh is crossing the politically volatile deltaic regions. Even though the original reason for constructing this fence was related to trafficking and smuggling, particularly cattle rustling, sea-level rise is providing a new incentive for its ongoing expansion as refugees flee into India from Bangladesh.

The Arctic

A world away from Bangladesh, the Arctic region has been described as a barometer for climate change. The Arctic and the Pacific island communities do, however, share some similarities. The people in both regions rely on passed-down cultural knowledge and natural resources for survival, retaining a connection to the environment through a body of traditional knowledge developed over centuries.

The ancestors of the indigenous people that inhabit western Alaska today likely arrived around 5,000 years ago. Alaska's migrants then were mobile subsistence hunter-gatherers, a lifestyle that remains a core part of their existence. But the inhabitants now are in jeopardy. The U.S. Government Accountability Office has concluded that of the more than 200 native Alaskan villages, 85 percent are already being affected by erosion and flooding, and 31 villages are under imminent threat of inundation. As we noted earlier, 12 Alaskan beachfront communities have decided to relocate completely. They have no choice.

Africa

The African continent is bounded in the north by the Mediterranean Sea, the Atlantic Ocean to the west, the Indian Ocean to the east and southeast, and the Red Sea to the northeast, connecting to the Mediterranean via the Suez Canal. Africa is composed of nearly

50 countries and 33 coastlines, and as it is for all other regions of the world, sea-level rise is expected to cause problems for infrastructure, transportation, agriculture, and water resources at the coast. Of the 136 port cities with asset and population exposure identified in the study for the OECD conducted by Robert Nicholls and his colleagues, 19 are in Africa. However, for Africa, the impacts of sea-level rise are poorly understood, and many African countries are not able to prepare, respond or adapt as needed. In a United Nations Environment Programme 2011 paper, Sally Brown and her coauthors studied the effects on Africa of sea-level rise and found that intense poverty, lack of historic studies and current data for making decisions, and strong trends toward urbanization all are factors in the potential impact expected for the continent.

The River Deltas

River deltas can be found throughout the world in both hemispheres, including at the mouths of the Mekong, Irrawaddy, Niger, Nile, Mississippi, Ganges–Brahmaputra, and Yangtze rivers. *Deltas* are highly dynamic landforms shaped by fluvial and coastal flooding. Dense populations crowd the deltas, primarily because they are flat and covered with rich agricultural soil to support life. These populations are at risk and will need to relocate.

A report by the IPCC listed Bangladesh, the Nile Delta in Egypt, and the Mekong Delta in Vietnam as the world's three "hot spots" for potential migration because of their combination of sea-level rise and existing populations. In 2009 Seth Mydans wrote in the *New York Times* that in a worst-case projection, more than one-third of the Mekong Delta, where 17 million people live, could be submerged if sea levels rose by 3 feet in the coming decades. In a more modest projection, one-fifth of the delta would be flooded, according to Tran Thuc, who leads Vietnam's National Institute for Hydrometeorology and Environmental Sciences and is the chief author of the report relied on by Mydans.

To put this issue of ethics and justice in perspective for Americans, Hurricane Katrina caused one of the largest and most abrupt displacements of people in U.S. history, with an estimated 1.5 million people hastily leaving their homes along the Gulf Coast. People across all demographics were forced to adapt to the environmental upset. Only some were able to return home, and those returning did not resemble the same income or demographics of those who left—54 percent of African American temporary evacuees returned, compared with 82 percent of white evacuees. Even though billions of dollars were paid to the victims of Hurricane Katrina, residents in nonwhite areas were paid less than were those living in mostly white communities, according to a 2010 Amnesty International report.

Using the word *refugee* to describe this environmental migration is controversial, as if somehow this strips dignity from people who were forced into a situation in which they never should have been put in the first place. The communities where Katrina refugees settled in large numbers—for instance, Houston and Atlanta—have been changed. It therefore is important that we have a framework for this status in order to provide services for those who are forced to migrate and to avoid such stigma in the future.

Clearly, environmental refugees from undeveloped countries face numerous problems. Herding and farming skills are not relevant in urban areas. Rural farmers are often more self-sufficient than many urban dwellers are, but they may not be used to depending on other people or a corporation for employment. Even though environmental refugees must adjust to different laws, languages, and cultures, they create many problems for their new countries. Educational and health-care systems must adjust to a new, suddenly appearing population that may speak a different language or whose customs differ from those of the native population.

What do we know, and what can we assume from lessons learned in the past? We do know that people have always moved, for various environmental and nonenvironmental reasons. That will continue, with sea-level rise being only one influence of many, but certainly a major one. One city on the forefront of the nonpolitical-refugee problem is Dhaka, Bangladesh, one of the world's fastest-growing cities. In 2005, its population was 12 million and now is more than 17 million. We also know that climate change is causing the entire planet to change rapidly in alarming ways. Finally, we know that the global human population has never been as large as it is today.

This looming crisis requires the world's attention because the cost of not acting now will be immeasurable. Environmental retreat can be carried out in an organized and voluntary way, but some people still oppose extending the definition of refugees. Moreover, solutions to the environmental refugee problem will not come easily because they must be global, as the current institutions, organizations, and funding mechanisms are not sufficiently equipped to deal with this problem.

10

Deny, Debate, and Delay

The debate is over. Don't believe us? Ask a climatologist. In a literature review, of those climatologists who expressed an opinion, 97 percent agreed that the climate is warming and that the warming trends of the past century are very likely caused by human activity. For example, a report in 2013 by the IPCC stated that there is a 95 percent certainty that most of the observed warming over the past century had been caused by human activity.

Interestingly, meteorologists are not nearly so certain about the cause of global climate change. But remember that dealing with day-to-day weather is different from studying climate on a decadal, century, or millennial scale. A survey published by the American Meteorological Society in 2013 found that its members who were less active researchers (as indicated by the publication of peer-reviewed papers) were more likely to be climate-change skeptics, just as those were who identified themselves as conservatives. In other words, those less expert in the field and those driven by conservative ideology are less likely to recognize climate change. Specifically, the survey found that 93 percent of the climate scientists who publish on climate change and other topics are convinced that humans have contributed to global warming, compared with 65 percent of nonpublishing climate scientists and 59 percent of nonpublishing meteorologists who do believe that humans have contributed to global warming.

According to a Pew Research Center poll conducted in 2013, 69 percent of Americans now believe there is solid evidence of global warming, but only 42 percent believe the warming is due mostly to human activities. A 2012 Pew poll showed that while 85 percent of Democrats said there was solid evidence of warming, only 48 percent of Republicans shared that view. What explains the gap between climatologists' science-based knowledge and the U.S. public's, particularly Republicans', belief? Some of the blame falls on the duplicitous efforts of the climate-change deniers' lobbyists, who are backed by the fossil-fuel industry. Some blame falls on the media for poor reporting and their apparent need to generate controversy rather than to report the science. And yes, some of the blame falls on the science community for not effectively "going public."

The roots of the campaign to deny the existence of climate change or to delay any significant action to combat climate change go back to the tobacco industry's campaign to deny the harmful nature of cigarettes. In fact, some of the same scientists and lobbying groups that managed to delay regulatory action on tobacco for decades, by downplaying the risks of cigarettes, have also helped delay regulatory action regarding climate change.

Frederick Seitz

The initial shots in the battle over climate change, the "Fort Sumter" if you will (coincidentally, a historical site also threatened by sea-level rise), came in the form of a report published by the George C. Marshall Institute (GMI). The GMI was initially formed to focus on defense issues but evolved into a leading contrarian climate-change think tank. In its 1989 report, the GMI did not dispute warming but argued that any warming was due to cyclical variations of the sun and that the cycle suggested the planet would actually cool in the twenty-first century.

In 1990, in its first assessment of climate change, the IPCC explicitly rejected the GMI's position blaming the sun. But the GMI had already briefed the Bush administration, and in a letter to a vice president of the American Petroleum Institute, coauthor Robert Jastrow bragged that the GMI was "responsible for the Administration's opposition to carbon taxes and restrictions on fossil fuel consumption."

One of the coauthors of the GMI paper, which was turned into the book *Global Warming: What Does the Science Tell Us?* was physicist Frederick Seitz. Seitz, a former president of the prestigious National Academy of Sciences (NAS), worked for R. J. Reynolds helping distribute its funding of medical research from 1978 to 1988. In a 1994 paper, once again published by the George C. Marshall Institute rather than in a peer-reviewed journal, Seitz not only expressed doubt, based on "sound scientific work," over whether we are in immediate danger from either global warming or depletion of the ozone layer, but also managed to work in the statement that there is "no good scientific evidence" that secondhand cigarette smoke "is truly dangerous under normal circumstances."

Arthur Robinson

In 1998, Arthur Robinson of the Oregon Institute of Science and Medicine circulated in a mass mailing what has become known as the "Oregon Petition." Attached to it was an unpublished paper, by Robinson with Willie Soon and Sallie Baliunas, that looked like a publication of the National Academy of Sciences. It used the same typeface and format as official NAS proceedings, along with a cover note signed by its former president, Frederick Seitz. The petition urged the U.S. government to reject the Kyoto Protocol or any similar proposals, stressing that the "proposed limits on greenhouse gasses would harm the environment, hinder the advance of science and technology, and damage the health and welfare of mankind." It further argued:

> There is no convincing scientific evidence that human release of . . . greenhouse gasses is causing or will, in the foreseeable future, cause catastrophic heating of the Earth's atmosphere and disruption of the Earth's climate. Moreover, there is substantial scientific evidence that increases in atmospheric carbon dioxide produce many beneficial effects upon the natural plant and animal environments of the Earth.

The petition quickly picked up 19,000 signatures (and was up to 31,487 by January 2010). The NAS did issue a statement stressing that the "petition does not reflect the conclusions of expert reports of the Academy." This petition offers an excellent example of the climate-change deniers' strategy. Express a contrarian view in a non-peer-reviewed format and then repeatedly promote the paper or, in this case, talk up the signatures on a misleading document. Not only had the paper bypassed the traditional peer-review process, but it was deceptively packaged to look like it was a peer-reviewed publication! David McCandless, a journalist and blogger, and Helen Lawson Williams examined the backgrounds of the signatories and concluded that 49 percent were engineers. We should note that contrarians often are scientists or professionals whose areas of expertise lie outside the field of climatology. Scientists who deny climate change and our role in it typically produce flawed papers with cherry-picked facts published in non-peer-reviewed journals. The results are then trumpeted by right-wing think tanks, websites, and news organizations like Fox News. These papers are eventually debunked in scientific journals a year or so later, to much less fanfare, but nonetheless they have served their purpose of creating doubt in the minds of the public.

Indeed, despite its deceptive format, the Oregon Petition served its purpose as a tool of doubt. For instance, it allowed Diane Bast of the Heartland Institute, a prominent think tank and source of climate-change-denier propaganda and funding, to proclaim that the debate over climate change was not settled. In an article on the institute's website, she stressed that more than 30,000 "American scientists reject the assertion that global warming has reached a crisis stage or is caused by human activity." With regard to this "debate," science historian Naomi Oreskes reviewed the abstracts of 928 papers on global climate change published in scientific journals between 1993 and 2003 and found not a single one that did not explicitly or implicitly accept the human role in climate change.

In their important 2010 book *Merchants of Doubt*, Oreskes and Erik Conway convincingly link Big Tobacco with efforts to discredit science harmful to corporate interests, not just those of Big Tobacco, but to other corporate interests such as the fossil-fuel industry. They took the title of their book from a letter from the tobacco industry proclaiming, "Doubt is our product since it is the best means of competing with the 'body of fact' that exists in the mind of the general public. It is also the best means of establishing a controversy." Oreskes and Conway detail the role of Frederick Seitz and others in the creation of a systematic approach to combating science and government regulation. The tobacco company Philip Morris may even have coined the term *junk science* for peer-reviewed studies that might harm its industry, as opposed to *sound science* for studies that support its views. Steven Milloy, operator of the junkscience.com website, has popularized the notion of junk science while making light of the threat from climate change. Once employed as a lobbyist for a firm hired by Philip Morris to downplay the dangers of secondhand smoke, Milloy was executive director of the Advancement for Sound Science Center (formerly Coalition) (TASSC), a group that, Oreskes and Conway state, is dedicated to discrediting, rather than advancing, science. A commentator for Fox News, Milloy is a well-positioned cog in the climate-change-denial machinery.

S. Fred Singer

TASSC is an industry-funded lobbying group formed by APCO Worldwide, a public-relations firm hired by Philip Morris. Initially created to manufacture doubt about the harmful effects of cigarettes, the group expanded to other issues, by design, in order to avoid seeming like a front for Big Tobacco. S. Fred Singer, a physicist, was one of TASSC's scientific advisers. He is known for founding the pro-tobacco Science & Environmental Policy Project and for producing publications rejecting the dangers of secondhand cigarette smoke, as well as for denying the human role in climate change while stressing that global warming will actually benefit humanity. Leaked documents from the Heartland Institute, a major climate-change-denier think tank, reveal that Singer receives $5,000 per month for work attacking the IPCC. Singer and fellow scientist-for-hire Craig Idso, who reportedly receives $11,600 per month through his Center for the Study of Carbon Dioxide and Global Change, created the Nongovernmental

International Panel on Climate Change (NIPCC), which the Heartland Institute funds to explicitly rebut the IPCC's reports.

A report by the Union of Concerned Scientists, "Smoke, Mirrors and Hot Air," reveals that ExxonMobil used tactics forged by Big Tobacco to create uncertainty over climate change:

- Questioning even indisputable scientific evidence showing their products to be hazardous
- Engaging in "information laundering" by using and even establishing seemingly independent front organizations to make the industry's case and confuse the public
- Promoting scientific spokespeople and investing in scientific research in an attempt to lend legitimacy to their public relations efforts
- Attempting to recast the debate by charging that the wholly legitimate health concerns raised about smoking were not based on "sound science"
- Cultivating close ties with government officials and members of Congress

The report further details how ExxonMobil used these techniques to create doubt in the public's mind about the scientific consensus regarding climate change.

Finally, in 2008 ExxonMobil bowed to public pressure and stopped openly funding the climate-change deniers. Since then, however, groups and individuals funding those seeking to deny the existence or the importance of climate change have increasingly turned to "dark money" or undisclosed financing. Robert Brulle of Drexel University found a rapid increase in the percentage of funding of what he deems the counter-climate-change movement by Donors Trust and Donors Capital, foundations that hide their contributors' identity. He pointed out that this increase in the use of "dark money," while the denier funding from the foundations supported by Charles and David Koch and ExxonMobil has waned, coincides with campaigns by the Union of Concerned Scientists and Greenpeace to publicize and criticize those two groups as funders of the climate-change deniers.

Willie Soon

Being a contrarian can be quite beneficial to one's career. Wei-Hock "Willie" Soon, a researcher with the Harvard Smithsonian Center for Astrophysics, knows firsthand of the glories that come with being a climate-change contrarian. There's the celebrity that comes with the occasional opportunity to appear on Fox News or before a congressional committee to counter the latest report confirming the existence of climate change. There are the backslapping "attaboys" from his cronies at the Heartland Institute, at whose climate-change conferences Soon is a regular speaker. And there's the money. Since 2002, fossil-fuel industry and conservative groups have been the sole source of Soon's funding. According to documents obtained by Greenpeace via the Freedom of Information Act, over the past dozen years Soon has received research funding of more than $1.2 million from sources such as ExxonMobil, the Koch brothers, the American Petroleum Institute, and the Southern Company, a utilities holding company that was found to be the largest greenhouse-gas emitter in the utilities industry in a 2007 study by the Center for Global Development. Some of these funds have flowed from the Donors Trust, the secretive funding route that between 2002 and 2010 allowed conservative billionaires like the Koch brothers to channel close to $120 million to more than 100 groups that cast doubt on the science of climate change or oppose environmental regulations. So the money is good, and it also must be nice for a fellow whose area of expertise is solar research to enjoy a celebrity status among conservatives. Publish a few articles on solar variability, and you might toil away in relative obscurity; but coauthor a piece challenging Michael Mann, Raymond Bradley, and Malcolm Hughes's "hockey stick" paper, published in *Nature* in 1998, and you will find yourself in great demand among conservative think tanks dedicated to creating doubt about climate change in the minds of the public. Mann's "hockey stick" graph shows that temperatures declined over the past millennium until they rapidly rebounded over the past 150 years. The graph has been disputed by the usual suspects (George C. Marshall Institute, Fred Singer, Willie Soon, and so on) but has generally been supported by subsequent temperature reconstructions.

Soon's career also sheds light on some of the tactics used by the fossil-fuel industry to deny climate change. Like the work of other climate-change contrarians, many of his articles are not peer reviewed, and they frequently appear in publications of climate-change skeptics like the George C. Marshall Institute or skeptic-friendly journals like *Energy and Environment*. The faulty scientific assertions from these papers then flow directly into the mouths of oil industry–supported politicians like Congressman Joe Barton (R-Tex.) and Senator James Inhofe (R-Okla.). Soon's major contribution to the climate-change "debate" was a couple of papers in 2003 in which he and frequent coauthor Sallie Baliunas asserted that the twentieth century was not the warmest century in the past 1,000 years. So flawed was this paper that 13 of the scientists cited in it offered a rebuttal asserting that the authors' claims that recent warming was not unprecedented in the past millennium was "inconsistent with the preponderances of the scientific evidence." Soon's papers downplayed climate change and suggested that natural variations were to blame for temperature changes. Despite widespread criticism of the papers, Soon's work was cited by Inhofe to deny the connection between human activity and global warming.

Soon has also published a "viewpoint" article in the journal *Ecological Complexity* asserting that polar bears are not endangered. Such an article may seem out of place for an astrophysicist, but not at all for a scientist whom the fossil-fuel industry has funded for the past decade. Perhaps there is something about industry money that makes a scientist comfortable with branching out beyond his or her area of expertise? In a February 21, 2015, article posted on the investigative news blog *Inside Climate News*, Kert Davies, executive director of the Climate Investigations Center, who along with Greenpeace released e-mails from Soon to fossil-fuel industry funders, was quoted as saying, "Climate change isn't something you can go out and see like a polluted river. But with the polar bears, all of a sudden you could see the results of climate change." Soon's paper cast doubt on studies showing that polar bears are seriously endangered by climate change destroying their habitat. Consequently, the fossil-fuel industries were able to use Soon's paper to counteract a possible public-relations disaster: beautiful endangered bears that could become the face of climate change.

In a case of the pot calling the kettle black, Soon, speaking at a senior center in Georgetown, Delaware, in 2013, pooh-poohed dire sea-level rise projections as conclusions based on "flawed data that are manipulated and agenda driven." A February 15, 2015, front-page article in the *New York Times* reveals that in e-mails and grant contracts between Soon and his fossil-fuel industry funders, Soon refers to the papers for which he is handsomely rewarded as "deliverables." The *Times* article stresses that in at least 11 papers published since 2008, Soon failed to disclose the conflict of interest in which he had been paid more than $1.2 million by the fossil-fuel industry in the past 10 years. Furthermore, the article states that in at least eight of those cases, Soon appeared to violate the ethical guidelines of the journals that published his work.

Nils-Axel Mörner

Nils-Axel Mörner is a colorful, retired Swedish professor who has also enjoyed some notoriety from espousing contrarian views of climate. Former president of the International Union for Quaternary Research (INQUA)'s Commission on Sea-Level Changes and Coastal Evolution, his skeptical viewpoint has been explicitly rejected by the INQUA itself, as indicated in a 2004 letter by the then president of INQUA, John Clague, in which he stated that nearly all INQUA researchers agree that humans are modifying the earth's climate.

Mörner is an old-school geologist in that he favors field observations over quantitative models. There is something to be said for this view, because models often are based on vague principles and limited data. He reportedly asserted that the claims about sea-level rise are 100 percent wrong because they are based on computer model predictions, whereas his findings are based on "going into the field to observe what is actually happening in the real world." Besides the ubiquitous evidence of sea-level rise from tide gauges and satellite measurements, which Mörner rejects in favor of his type of field observations, a host of other indicators (from the "field") confirm that the sea level is rising. For example, when the moon is full (spring tide), people in parts of Miami Beach must drive through saltwater on some roads, as do the citizens of Norfolk, Virginia, and parts of Annapolis, Maryland. These are roads that 40 years ago were flooded only in storms. Parts of the Pamlico and Albemarle Sounds as well as the edges of marshes on Assateague Island, Virginia, are rimmed with dead trees, killed by brackish waters reaching higher levels upstream as the sea level rises. Salinization caused by a rising sea is evident in many of the world's atolls where the freshwater lenses no longer exist or are very thin and people must depend on rainwater collected in cisterns. One well-known example is seen in a set of photographs from 1949 and 1981, first published by University of Miami geologist Hal Wanless, taken of a bridge abutment in Miami showing that the barnacles and oysters have moved upward, demonstrating a rise in sea level. Many other small examples show a change in sea level, such as old cemeteries in the salt marshes of Ocracoke Island, North Carolina, where they never would have been placed originally. In addition, physics requires that ocean water expand as it heats up. That is, we have measured increases in the temperature of the water in the uppermost 2,000 feet of the ocean, which we know, through lab experiments, should cause a sea-level rise through the expansion of water. So the field observation of increased ocean temperatures must result in expansion and a rise in sea level.

Mörner cites his research in the Maldives as proof that sea-level rise is overblown. He insists that evidence from his field observations shows that around 1,000 to 800 years ago, the sea level was approximately 2 feet higher than it is now and that there has been a recent 1-foot drop in sea-level rise in the Maldives, most likely occurring in the 1970s. Mörner cites, in part, the discovery of a woman's skeleton found in the coral reef for his belief that seas were higher in the past, and uses anecdotal evidence from fishermen to support his assertion that the sea level recently fell by 1 foot. His findings were discredited by Philip Woodworth of the United Kingdom's National Oceanography Centre in a paper published in the same journal in which Mörner made his assertions. In a December 2011 article in the *Spectator*,

Mörner also discounted satellite altimetry data showing as untrue a sea-level rise of around 0.12 inch per year since 1993, suggesting that the data had been "hijacked and distorted by the IPCC for political ends." One strange assertion was his false claim that the rate of sea-level rise supported by most climate scientists was based on a single tide gauge in Hong Kong.

Despite his discredited Maldives claims, Mörner reasserted them in the aforementioned *Spectator* article. In so doing, he described INQUA as having the world's experts in sea-level rise and invoked his status as a past president, despite the organization's explicit rejection of his decidedly minority view. In its "Statement on Climate Change," INQUA maintains that there is strong evidence of significant global warming, citing evidence from "direct measurements of rising surface air temperatures and subsurface ocean temperatures and, indirectly, from increases in average global sea levels."

Mörner and Soon teamed up for a 2012 op-ed in the *Washington Times* in which they attempted to discredit the reliability of tide gauges by citing a rather creative study by engineer Cyrus Galvin. Galvin attributes the fluctuations in the tide records from Atlantic City's Steel Pier to the weight of spectators gathered to watch horses ridden bareback by women dive off the pier into a pool, videos of which are available on YouTube. Thousands would gather to watch the stunt dives, resulting in loading and offloading of tons during the multiple daily shows, almost like a pile driver. Soon and Mörner note a slowdown in the rate of sea-level rise on the pier's tide gauge from 1945 through 1953, a period during which the horse diving was suspended, thus proving that the diving horse was responsible for the observed sea-level changes. The likelihood of this thesis plummets, like a mustang from a diving board, because an examination of other tide-gauge data from the eastern United States also shows a similar drop in the rate of sea-level rise in the same time frame, apparently due to natural factors. Another factor involved is the variable resistance of the substrate into which the pier was being driven. The biggest problem with Mörner and Soon's claim is that their assertions are based on a single data point. One data point is a useless piece of information concerning a global phenomenon.

In a paper published in *Theoretical and Applied Climatology* in 2015, Rasmus Benestad and his colleagues set out to replicate the studies of climate contrarians and found some common methodological shortcomings. Coauthor Dana Nuccitelli, writing in the *Guardian*, noted that the studies shared such methodological flaws as cherry-picking facts, ignoring important contextual information or important data that conflicted with research conclusions, disregarding known physics, and relying on curve fitting. Nuccitelli pointed to a 2011 paper by Ole Humlum, Jan-Erik Solheim, and Kjell Stordahl stating that the authors had discarded a good bit of data from the Holocene era that did not fit their claims—data that, if included, would negate the predictive outcomes of their model. Nuccitelli quoted mathematician John von Neumann on the practice of curve fitting as "with four parameters I can fit an elephant, and with five I can make him wiggle his trunk."

A Success Story in Denying Sea-Level Rise

In 2012, the sea-level debate was successfully hijacked in North Carolina. The advisory Science Panel for North Carolina's Coastal Resources Commission was charged with preparing "a report, based on a review of the published literature, of the known state of sea-level rise for North Carolina." This report was to include sea-level rise projections through 2100. The panel concluded that by 2100, a 15-inch rise was certain, a 39-inch rise was likely, and a 55-inch rise was possible. The panel thus recommended that the 39-inch projection be adopted as the basis for future coastal management.

This report unleashed an antiregulatory backlash led by NC-20, a group led by business and local government individuals from 20 North Carolina coastal counties. Realizing that a projection of a 39-inch rise in sea level would result in costly regulations regarding development, NC-20 launched a lobbying and public-relations effort to defeat regulations based on projections of sea-level rise. It succeeded in at least delaying regulations that, in the long run, would save both money and lives for the constituencies of these 20 counties.

Through lobbying efforts that included public presentations from climate-change deniers, NC-20 persuaded the Coastal Resources Commission to abandon the 39-inch projection. State representative and real-estate broker Pat McElraft cosponsored N.C. House Bill 819, legislation stipulating that only "historic[al] rates of sea-level rise may be extrapolated to estimate future rates of rise but shall not include scenarios of accelerated rates of sea-level rise unless such rates are from statistically significant, peer-reviewed data and are consistent with historic[al] trends." In effect, this bill would have resulted in projecting a sea-level rise of around 8 inches by 2100. One might wonder why quibbling over 39 inches could result in such controversy, but because North Carolina's coastal plain has a very gentle slope, a rise of 39 inches could push the shoreline back roughly 2 to 4 miles. This possibility is why NOAA recognizes northeastern North Carolina as one of the three regions in the United States most susceptible to sea-level rise, with the Mississippi Delta and South Florida being the others.

In 2012, television comedian Stephen Colbert praised (tongue-in-cheek) the North Carolina legislature for having solved the problem of facing unpleasant scientific reality. If you don't like the conclusions, such as climate models showing a future of sea-level rise, just legislate that it won't happen! Faced with models predicting a devastating sea-level rise of 39 inches, the legislature limited predictions to those based on projected historical data (rather than mathematical models), which lowered the prediction of sea-level rise to just 8 inches. That a legislature would even attempt to set such a denial into law should leave the electorate breathless.

Ultimately, the state neither passed a law that would make sea-level rise illegal nor limited projections of sea-level rise solely to those based on historical data. Instead, the law ignored the recommendations of the Science Panel's report by not calling for consideration of sea-level rise in coastal planning. In addition, the law called for a 2015 sea-level rise report that included a "summary of peer-reviewed scientific literature that address[es] the full range of global, regional and North Carolina-specific sea-level-change data and hypotheses, including sea-level fall, no movement in sea level, deceleration of sea-level rise, and acceleration of sea-level rise." This wording sets up another battle over the science by explicitly calling for the inclusion of the types of papers cited by NC-20 in its opposition to the Science Panel—papers that have been discredited by the scientific community.

How did NC-20 succeed in derailing regulations that consider the fact that the sea level is rising? For one thing, it was highly motivated by financial self-interest. But let's keep in mind that by defeating a reasonable first step toward new regulations that would take into account a rise in sea level, NC-20 not only has cost the taxpayers more money because of the expense of recovering from flooding, but also will cost lives because of the delay of regulations that will make coastal communities more resilient.

Tom Thompson, chairman of NC-20 and the economic development director of Beaufort County, explained that NC-20 members are convinced that climate changes and sea-level rises are part of natural cycles. Thompson's statement that climate change is part of a natural cycle will by now sound familiar to those acquainted with the arguments spun from the climate-change deniers. NC-20 has used numerous tactics employed by the deniers to defeat the regulations. For example, it portrayed the Science Panel, most of whose members are internationally renowned in their fields of oceanographic, geologic, and engineering specialties, as extremists who are motivated to push a climate-change agenda to ensure being granted funding. NC-20's experts also argued there is no scientific consensus on sea-level rise, even though it is generally regarded as one of the more predictable outcomes of climate change. They cherry-picked both data and experts, relying heavily on Nils-Axel Mörner's work, which, as we noted earlier, has been totally discredited.

At the time of this writing, the North Carolina Science Panel members had issued a 2015 draft report, with the final report due in spring 2016. The state asked for estimates for only the next 30 years, the time span over which the scientific community anticipates a slow, gradual rise in sea level, to be followed thereafter by a rapid rise. The prediction's 30-year span was chosen in order to avoid the frightening 100-year span prediction.

The science panel offered sea-level rise predictions for North Carolina based on current rates of sea-level rise, a projected sea-level rise with low greenhouse-gas emissions, and one with high greenhouse-gas emissions. In the scenario with low greenhouse-gas emissions, they projected a sea-level rise ranging from a low estimate of 5.8 inches in Wilmington to 7.1 inches in Duck. In the scenario with high greenhouse-gas emissions, the projected sea-level rise over a 30-year period varied from a low estimate of 6.8 inches in Wilmington to a high estimate of 8.1 inches in Duck. In general, higher rates of sea-level rise were predicted for northeastern North Carolina, owing to the higher rates of subsidence in that region.

A September 23, 2014, e-mail from David Burton, a member of the NC-20 and climate-change contrarian, to the members of the Science Panel called for "balancing the information sources" and is revelatory. Burton describes the National Climate Assessment and the IPCC's fifth Assessment Report as coming from the "left" end of the spectrum. To balance out these sources, he suggested that the committee consider reports from the Nongovernmental International Panel on Climate Change (discussed earlier) and the Committee for a Constructive Tomorrow (CFACT), a conservative think tank whose board of advisers has included such well-known contrarians as Sallie Baliunas, Patrick Michaels, and the late Frederick Seitz. Burton also suggested that the board rely on information from the conservative climate contrarian website Wattsupwiththat.com run by former television meteorologist Anthony Watts. A fair and equal balance is difficult to achieve if one side of the scales is piled high with horse manure. Yet this is the task faced by scientific panels or policymakers making crucial life-and-death decisions, a task made all the more difficult by a climate-change countermovement funded by energy interests. For North Carolina, sea-level rise will be the devastating arm of global warming. As long as the state's citizens fall in line behind a legislature and lobbying groups denying climate change and the accelerating sea-level rise, the price will be paid in more and more lost property and lives.

A fair and equal balance is difficult to achieve if one side of the scales is piled high with horse manure.

The following are the lessons and facts that the politicians in North Carolina (and other politicians as well) must learn if the deniers are to be disarmed:

- Climate change is not a "belief": the science tells us it is real.
- Voting on whether you believe in climate change does not change reality.
- Human activities are the main cause of the recent climate change: the science tells us so.
- The current change in climate is not due to "natural cycles": the science tells us so.
- The rate of sea-level rise is accelerating: the science tells us it is real.
- Legislation cannot change the progression of nature: the rise in sea level will continue.
- We can plan endlessly, but actions are needed now to prepare for what is coming.

11

Ghosts of the Past, Promise of the Future

What do the British seaside towns of Orwithfleet, Sunthorpe, Ravenser Odd, Sisterkirke, Hornsea Beck, Hornsea Burton, and Wilsthorpe have in common? Answer: They no longer exist. All now reside on the seafloor off the Holderness Coast of Yorkshire in the United Kingdom. All were once coastal villages, and all were lost to the sea between the Roman invasion of Britain and the twenty-first century. The rise in sea level was slight during this time interval and thus was not a major cause of the changes in the shoreline.

What the Holderness experience tells us is that the behavior of the world's shorelines is a continuous process, which, in the very long view, cannot be halted except at great economic and environmental cost. As the rate of sea-level rise increases to the point that it is no longer viable to remain near the coast, the Holderness experience becomes a strong argument for the retreat option.

The Holderness Coast

The Holderness Coast is a stretch of shoreline of about 45 miles extending from Flamborough Head, a cape underlain by hard rock, southeast to Spurn Point, a sand spit at the mouth of the River Humber estuary (figure 3). Between the cape and the river, the coast consists of bluffs that are made up of unconsolidated glacial till

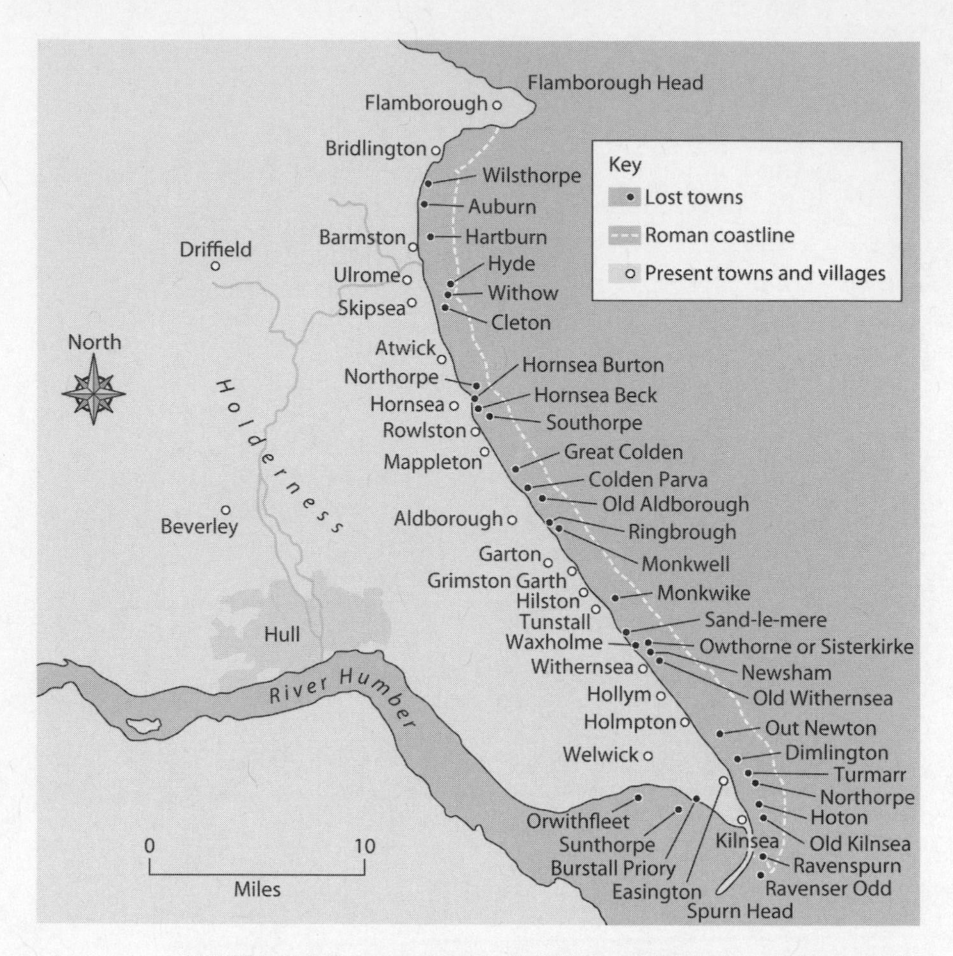

FIGURE 3 The lost towns of the Holderness Coast of Yorkshire, England. Since the Roman invasion, 28 villages have been inundated by the eroding shoreline. (After Thomas Sheppard, *The Lost Towns of the Yorkshire Coast and Other Chapters Bearing Upon the Geography of the District* [London: Brown, 1912])

(locally called boulder clay) and range from 10 to 115 feet in height. The till is composed in large part of mud and sand with occasional boulders, deposited during the last glacial advance. The sand and gravel that erode out of the till are the source of sediment for the adjacent beaches. The coast faces a long northeasterly fetch, which is responsible for high storm waves from the North Sea, causing rapid erosion rates of the "soft" bluffs. The Holderness Coast has among the highest overall erosion rates in all of Europe. In 1912, in *The Lost Towns of the Yorkshire Coast*, Thomas Sheppard suggested 9 feet per year as a good average shoreline retreat rate figure at that time.

The Holderness Coast is a macrotidal coast—that is, a coast with high tides, which here can be as high as 23 feet. Such a large tidal range means that a storm striking the shore at high tide will do much more damage than the same storm striking the coast at low tide. Because the inner continental shelf (the shore face) is steep, there is relatively little friction between the incoming waves and the seafloor. Thus, North Sea waves directly strike the coast with their full energy, and at high tide, in most areas the waves directly strike the base of the bluffs. Extreme annual wave height (the highest wave expected in a year) is on the order of 15 feet. The extreme wave height expected for a 100-year storm recurrence interval is 23 feet. Add to this the sea-level rise, which will be essentially the same as the global sea-level rise, plus the fact that this coast is also experiencing land subsidence, and the future of this coast should be clear. The latter subsidence alone, which is expected to be 3.5 inches annually by 2050, is sufficient to guarantee more bluff recession during storms. Overall, the intermediate prediction for sea-level rise along the Holderness Coast is 1 foot by the year 2050.

The erosion of bluffs like those of the Holderness Coast is a complex process involving a number of simultaneously operating processes, which include

- *Abrasion* by rocks thrown against the bluffs by the waves
- *Attrition* of sand and rocks thrown against one another
- *Hydraulic action* by air and water in fractures being compressed
- *Dissolution* of rocks such as limestone
- *Wave pounding* that breaks off rock
- *Subaerial processes* such as rain, freezing and thawing, and chemical weathering
- *Downslope movement* of bluff material to the beach caused by gravity (e.g., creep, slumps, slides)

- *Biological processes* like those by burrowing organisms and tree roots that loosen bluffs
- *Human activity* such as water added to the bluff by septic tanks or lawn watering, and loading bluff edges with buildings and roads
- *Steep slopes offshore* allowing bigger waves than would a wide, flatter shelf that dissipates wave energy through friction between the waves and the seafloor

Ironically, another significant cause of erosion is coastal-engineering structures, especially groins, seawalls, and rock revetments that are intended to slow erosion rates at a particular location. The problem is that in the long run, each type of hard structure will cause downdrift erosion. This effect, caused by the cutting off of the sediment supply, is especially true of groins. About 16 percent of the shoreline of the Holderness Coast is armored. Some communities have both groins and seawalls (Hornsea), and some have both seawalls and boulder revetments on the same beach (Mappleton and Withernsea).

In *The Lost Towns of the Yorkshire Coast*, Sheppard states that the Holderness Coast has been lined with small towns and villages since the time of the Roman invasion in 43 C.E. In the 2,000 years since, the bluffed shoreline has retreated between 2 and 3.5 miles. All told, 28 small towns were abandoned, eroded into the sea, or drowned, and now lie on the seafloor. In all cases, the inhabitants retreated from the shoreline, so retreat is not a new solution to a rising sea level and an eroding coastline, but one that historically is known to work.

One of the more important lost villages was Ravenser Odd (Viking for "Raven's Tongue") at the southern end of the Holderness Coast near the mouth of the River Humber. The town was founded in 1235 and, by the early fourteenth century, had become a thriving seaport with many cargo ships and fishing boats. The town featured a chapel, a courthouse, and a prison, as well as a customs house and warehouses, and had two members in Parliament in 1295, indicating that it was both sizable and important. Ravenser Odd was just south of

Ravenspurn, which is referred to as Ravenspurgh in works by William Shakespeare:

> For all the world
> As thou art to this hour was Richard then
> When I from France set foot at Ravenspurgh;
> And even as I was then is Percy now.
>
> HENRY IV IN *HENRY IV, PART 1*, ACT 3, SCENE 2

> Wipe off the dust that hides our scepter's gilt
> And make high majesty look like itself,
> Away with me in post to Ravenspurgh;
> But if you faint, as fearing to do so,
> Stay and be secret, and myself will go.
>
> NORTHUMBERLAND IN *RICHARD II*, ACT 2, SCENE 1

Erosion of Ravenser Odd was a major continuing problem, and by 1346, most of the city had been damaged or destroyed. In 1355, bodies that had washed out of the cemetery near the chapel were reburied in a churchyard in Easington. The end came when the winter storms in 1356/1357 flooded the entire town. Everything that remained was later destroyed by the Grote Mandrenke storm of 1362. The 130-year-old town was gone forever.

All the world's ocean-facing coasts are threatened by a variety of forces, including nor'easters, king tides, hurricanes (cyclones, typhoons), winter, and El Niño storms, aided and abetted in the erosion process by the current gradual rise in sea level. All these events will push the world's shoreline back at an accelerating pace. But as geologist Gary Griggs pointed out, "It will be the extreme events [storms] that will pose the greatest risks to coastal development during at least the next several decades." The extreme events were also what caused the shoreline's continuing migration that overtook and destroyed the villages on the Holderness Coast.

Regarding extreme events, a storm in the North Sea in the winter of 1953 that struck at spring tide was one of the most devastating storms ever for the east coast of Britain (and did even more damage in the Netherlands, as explained in chapter 4). From a global standpoint, six of the world's "greatest storms" (each of which killed more than 100,000 people) struck Bangladesh. Examples of extreme storm events in North America include the Galveston hurricane in 1900, the Labor Day hurricane in the Florida Keys in 1935, the hurricane of 1938 in New England, Hurricane Hazel in 1957 in the Carolinas, the Ash Wednesday storm of 1962 in the mid-Atlantic, the El Niños on the California coast in 1982/1983 and 1997/1998, Hurricane Katrina in 2005 in the Gulf of Mexico, and Hurricane Sandy in 2012 on the U.S. east coast.

As sure as the sun will rise tomorrow morning, these extreme events will be repeated. But the results will be different because the population affected will be much larger, more developments will be in their path, and the rise in sea level will produce larger storm surges. In addition, the storms are expected to be more energetic owing to the warmer ocean waters.

As sure as the sun will rise tomorrow morning, these extreme events will be repeated. But the results will be different because the population affected will be much larger, more developments will be in their path, and the rise in sea level will produce larger storm surges.

One interesting observation reported by Sheppard was a process he termed *deep sea erosion*. That is, remnants of some ancient towns and villages are situated at what is now the 5-fathom (30 feet) line, 2 miles out from the current shoreline. These examples are a clear indication that given enough time, shoreline erosion is not just the landward movement of the wet–dry line on the beach. Shoreline erosion is also the landward movement of the entire nearshore system, including the shore face. Indications are that most of the eroded sediment from the bluffs is transported directly offshore. On the Holderness Coast, a small percentage is carried by long-shore currents to the mouth of the River Humber where a sand spit (Spurn Head) is maintained.

There was and still is an underlying erosion rate ranging from 3 to 10 feet per year and averaging 5 feet, but the big storms, the extreme events, were the final straw that destroyed buildings and forced the villagers to retreat. The annual underlying erosion rate appeared (at least in the past few centuries) to be handled by seawalls often constructed from the debris of storm-wrecked buildings. Sooner or later, the truly massive nor'easters from the North Sea did their job, and the towns' residents abandoned the wreckage of their homes and left for higher, safer ground. We see this process repeated today in New York and New Jersey, where some storm-fatigued residents eagerly accepted buyouts from the states in order to relocate away from neighborhoods repeatedly damaged by recent storms.

Storms will forever provide a reservoir of spectacular tales, some inevitably embellished over the years. For example, Sheppard tells the story of an old fisherman who lived in Withernsea:

> There were two houses between the old fellow's cottage and the crumbling cliff-edge. One rough night however a biting nor'easter hurled the ramping breakers against the shore to such purpose first one house went, and then the other. Then the wall of the old fisherman's cottage collapsed . . . and he awoke in the grey of the morning to find himself looking straight from his bed on to the green waters of the North Sea.

As the shoreline moved inexorably landward, treasures from churches and monuments often were rescued and moved back. A monument at Ravenspurgh was erected to honor the landing in 1399 of Henry IV, who was en route to dethroning Richard II. The monument overlooking the sea was forced to be moved, on unknown dates, first to Kilnsea, next to Burton Constable, and finally to Hedon.

Today, a thousand stories can be told about the Holderness bluffs. A reflection of the desperate days of World War II is the numerous pillboxes tumbling about on the beaches at the base of the bluffs. At Ringborough, concrete World War II pillboxes, a brick watchtower, and an artillery battery reside together on the beach, mixed with fragments of buildings from the original town. Some of the wartime concrete structures have broken up and have become rounded, flat boulders in the surf zone, slowly grinding down on their way to becoming sand grains on the beach. Across the sea on France's western shores, these ruins are mirrored by the ruins of German defenses, now fallen on the beaches or offshore due to beach migration.

On a bluff in the little English town of Easington is a plant that processes natural gas piped in from Norway. Roughly one-quarter of the gas required by the United Kingdom goes through this plant to be piped to users in various communities. It would seem to be a very stupid decision to build such a plant near the edge of an eroding bluff, but the original source of gas from offshore British fields was expected to be depleted 25 years after production began. But predicting the future is always dangerous, and no one guessed that British gas would be replaced with gas from Norway. Suddenly the plant's life span had to be lengthened. Accordingly, a massive rock revetment was built at the base of the threatened bluff, along with other measures that *so far* have saved the plant.

What the Holderness Coast Tells Us

The story of the Holderness Coast is one of coastal evolution on a millennial scale, a scale that doesn't directly concern us. But the 28 drowned villages on the continental shelf remind us that coastal change is continuous and forever. They remind us that nature bats last at the shoreline. Much of the beachfront development along America's shoreline, as well as that of other countries, is threatened by erosion at such rates that they will be gone in 100 to 200 years if nothing is done to stop the shoreline retreat. Moreover, this is before the expected acceleration of sea-level rise sets in. The future is already here for millions of coastal dwellers, and an immediate retreat is already being forced on the coastal inhabitants of coral atoll islands; Arctic shorelines; deltas, including the Mississippi, Ganges, Irrawaddy, Yangtze, and Niger deltas; and various shorelines such as the Holderness Coast. A global view of beachfront development, however, reveals little apparent concern among developers about a rising sea or the likely increased magnitude of storm surges and the inevitable need for a retreat from the shoreline. Miami, a city without a long-term future, is perhaps the biggest violator of common sense regarding the rising sea level. It is a city where officialdom and developers alike are whistling past the graveyard when it comes to climate change. Other examples (out of many) of major beachfront developments either planned or under construction are Oasis de Baños in Costa Blanca, Spain; Cable Beach in Nassau, Bahamas; Rosewood Jumbo Bay in Antigua; Palm Beach in Barbados; and Hard Rock Hotel in Cabo San Lucas, Mexico.

The future is already here for millions of coastal dwellers, and an immediate retreat is already being forced on the coastal inhabitants of coral atoll islands; Arctic shorelines; deltas, including the Mississippi, Ganges, Irrawaddy, Yangtze, and Niger deltas; and various shorelines such as the Holderness Coast.

One would think that the Maldives would be a different story. Here the government is aware of and publicizing the impact of sea-level rise on low-lying islands. While Mohamed Nasheed was president of the Maldives, he was a strong supporter of global change science and especially of research on sea-level rise. To make his point about sea-level rise, in 2009 he convened an underwater meeting of his cabinet, all fitted out in scuba gear, at a depth of 20 feet. Despite his recognition of the problem, the country, now under a new president, continues to encourage development. Nine new airstrips have been or are being constructed, and some of the islands are being sold to developers.

Is There Good News?

Indeed, there is good news. From one standpoint, we are well positioned to respond to the rising sea. As a society, we understand quite well how beaches and coasts operate. We know what a rising sea will do to the position of the shoreline and how different types of coasts will respond in their own particular way. We monitor sea-level rise through satellite and tide gauge measurements. We keep our eyes on the source of sea-level rise from the melting ice sheets and the warming oceans.

We understand the mechanics of coastal-engineering structures, seawalls, and groins, and we know the pluses and minuses of their emplacement on various shoreline types. We have learned that beach replenishment is costly, only temporary, and will become increasingly less feasible as the sea rises and the life spans of beaches grow shorter. Seawalls will hold the shoreline still for a while but at the cost of the beach's destruction.

Those who deny sea-level rise and climate change, though still with us in large numbers, are gradually being scuttled by a vastly increasing amount of evidence from many directions and places that point to this global phenomenon as being related to emissions of greenhouse gases.

Many states and countries now have regulations that promote light and well-placed development situated above storm-surge levels, though these regulations may not always be enforced.

Some steps toward retreat from coasts have been taken, such as the buyout approach being applied in locations as diverse as Staten Island, New York, and Galveston, Texas. FEMA is revising its flood maps to require building standards that are more stringent than the current ones. The Obama administration has mandated that future federal projects such as roads and sewage plants be constructed at least 2 feet higher than at present. Federal flood-insurance rates are gradually being raised to eventually reflect the full flood risk to buildings on the coast. While no communities are yet fully committed to the inevitable future retreat, our military is leading the way into a global-warming future as it plans for, among other things, a massive environmental refugee problem caused by sea-level rise.

There is not the slightest doubt that beachfront development will retreat on a massive scale, though widespread recognition of this and serious planning for it are lacking. In the meantime, until the problem becomes so obvious that even the most dedicated denier must give in, more local actions can be taken. First and foremost, building density should not increase, and large buildings (high-rises) must be prohibited. Good planning could include preserving space on the mainland to which buildings could be moved. New roads and other infrastructure should be placed as high and as far away from the shoreline as feasible. Disincentives to expand or stay in place must be applied.

The cautionary tales of the Holderness Coast, the lost islands of the Chesapeake, the lost neighborhoods of California, and the nineteenth-century ghost towns of the Carolinas must be heeded. The sooner we recognize the truth about nature's intentions at the shoreline, the better. Neither time nor tide is in our favor.

Bibliography

References

1. Control + Alt + Retreat

Andersen, Christine F., Jurjen A. Battjes, David E. Daniel, Billy Edge, William Espey Jr., Robert B. Gilbert, Thomas L. Jackson, and David Kennedy. "The New Orleans Hurricane Protection System: What Went Wrong and Why: A Report by ASCE Hurricane Katrina External Review Panel." Reston, Va.: American Society of Civil Engineers, 2007. doi: http://dx.doi.org/10.1061/9780784408933 (accessed September 4, 2015).

Barkham, Patrick. "Waves of Destruction." *Guardian*, April 17, 2008. http://www.guardian.co.uk/environment/2008/apr/17/flooding.climatechange/ (accessed May 1, 2008).

Bender, Hannah. "10 States at Greatest Risk of Storm Surge Damage." *Property Casualty 360°*, June 6, 2013. http://www.propertycasualty360.com/2013/06/06/10-states-at-greatest-risk-of-storm-surge-damage (accessed September 4, 2015).

Chang, Alicia, and Jason Dearen. "Beach Communities Moving Inward: Some Beach Towns Are Eyeing Retreat from Sea as Conditions Change with Global Warming." *Huffington Post*, June 2, 2012. http://www.huffingtonpost.com/2012/06/02/beach-communities-moving-inward_n_1565122.html (accessed September 4, 2015).

Davis, Richard A., Jr. *Beaches in Space and Time: A Global Look at the Beach Environment and How We Use It*. Sarasota, Fla.: Pineapple Press, 2015.

Esteves, Luciana S. *Managed Realignment: A Viable Long-Term Coastal Management Strategy?* Springer Briefs in Environmental Science. London: Springer, 2014.

Funk, McKenzie. *Windfall: The Booming Business of Global Warming*. New York: Penguin Press, 2014.

Gardiner, Stephen M. *A Perfect Moral Storm: The Ethical Tragedy of Climate Change*. Oxford: Oxford University Press, 2011.

Gornitz, Vivien. *Rising Seas: Past, Present, Future*. New York: Columbia University Press, 2013.

Government Accountability Office. "Alaska Native Villages: Limited Progress Has Been Made on Relocating Villages Threatened by Flooding and Erosion." Report GAO 09-551 to Congressional Requesters. June 2009. http://www.gao.gov/new.items/d09551.pdf (accessed September 4, 2015).

Griggs, Gary B. "Lost Neighborhoods of the California Coast." *Journal of Coastal Research Online*, November 18, 2013. http://dx.doi.org/10.2112/13A-00007.1 (accessed September 4, 2015).

Haddow, George, Jane A. Bullock, and Kim Haddow. *Global Warming, Natural Hazards, and Emergency Management*. CRC Press (online), 2008.

Johnson, Terrell. "11 Industries Poised to Profit from Global Warming." April 4, 2014. Weather.com: Environment. http://www.weather.com/science/environment/news/10-industries-are-going-get-rich-global-warming-20140402?pageno=3#/1 (accessed September 4, 2015).

Kelley, Joseph T., Orrin H. Pilkey, and J. Andrew G. Cooper, eds. "America's Most Vulnerable Coastal Communities." Geological Society of America Special Paper no. 460, 2009.

Klein, Naomi. "Don't Look Away Now, the Climate Crisis Needs You." *Guardian*, March 6, 2015. http://www.theguardian.com/environment/2015/mar/06/dont-look-away-now-the-climate-crisis-needs-you (accessed September 4, 2015).

LaVorgna, Marc, and Lauren Passalacqua. "Mayor Bloomberg Presents the City's Long-Term Plan to Further Prepare for the Impacts of a Changing Climate." June 11, 2013. City of New York,

Office of the Mayor. http://www1.nyc.gov/office-of-the-mayor/news/200-13/mayor-bloomberg-presents-city-s-long-term-plan-further-prepare-the-impacts-a-changing (accessed September 4, 2015).

Levermann, Anders. "The Inevitability of Sea Level Rise." *RealClimate*, August 15, 2013. http://www.realclimate.org/index.php/archives/2013/08/the-inevitability-of-sea-level-rise/#more-15633 (accessed September 4, 2015).

Myers, Norman. "Environmental Refugees, an Emergent Security Issue." Paper presented at the Thirteenth Economic Forum, May 23–27, 2005, Prague, Czech Republic. http://www.osce.org/eea/14851?download=true. (accessed September 4, 2015). Based on Myers, N., and J. Kent. *Environmental Exodus: An Emergent Crisis in the Global Arena*. Washington, D.C.: Climate Institute; and Myers, N. "Environmental Refugees: Our Latest Understanding." *Philosophical Transactions of the Royal Society B* 356 (1995): 16.1–16.5.

National Hurricane Center. "Storm Surge Overview." 2013; last modified September 5, 2014. Miami: National Weather Service. http://www.nhc.noaa.gov/surge/. (accessed August 28, 2015).

Nelson, Deborah J., Duff Wilson, Ryan McNeill, Alister Doyle, and Bill Tarrant. "Water's Edge: The Crisis of Rising Sea Levels." September 4, 2014. Thomsonreuters.com. http://www.reuters.com/investigates/special-report/waters-edge-the-crisis-of-rising-sea-levels/ (accessed September 18, 2015).

Nicholls, R. J., S. Hanson, Celine Herweijer, Nicola Patmore, Stéphane Hallegatte, Jan Corfee-Morlot, Jean Château, and Robert Muir-Wood. "Ranking Port Cities with High Exposure and Vulnerability to Climate Extremes: Exposure Estimates." OECD Environment Working Papers, no. 1. Paris: OECD, November 19, 2008. http://dx.doi.org/10.1787/011766488208 (accessed September 4, 2015).

Oreskes, Naomi, and Erik M. Conway. *Merchants of Doubt: How a Handful of Scientists Obscured the Truth on Issues from Tobacco Smoke to Global Warming*. New York: Bloomsbury Press, 2010.

Pilkey, Orrin H., William J. Neal, Joseph T. Kelley, and J. Andrew G. Cooper. *The World's Beaches: A Global Guide to the Science of the Shoreline*. Berkeley: University of California Press, 2011.

Pilkey, Orrin H., and Keith C. Pilkey. *Global Climate Change: A Primer*. Durham, N.C.: Duke University Press, 2011.

Pilkey, Orrin H., and Rob Young. *The Rising Sea*. Washington, D.C.: Island Press, 2009.

Siders, Anne. "Managed Coastal Retreat: A Legal Handbook on Shifting Development Away from Vulnerable Areas." October 2013. Columbia Center for Climate Change Law, Columbia Law School. http://web.law.columbia.edu/sites/default/files/microsites/climate-change/files/Publications/Fellows/ManagedCoastalRetreat_FINAL_Oct%2030.pdf (accessed September 4, 2015).

Strauss, Ben, and Remik Ziemlinski. "Sea Level Rise Threats to Energy Infrastructure: A Surging Seas Brief Report." April 19, 2012. Climate Central. http://slr.s3.amazonaws.com/SLR-Threats-to-Energy-Infrastructure.pdf (accessed September 4, 2015).

2. The Overflowing Ocean

Bruun, Per. "Sea Level Rise as a Cause of Shoreline Erosion." *Proceedings of the American Society of Civil Engineers; Journal of the Waterways and Harbors Division* 88 (1962): 117–30.

Clark, Pilita. "What Climate Scientists Talk About Now: As the Intergovernmental Panel on Climate Change Prepares to Release Its Latest Report, Pilita Clark Meets Some of the Key Scientists Behind It." *FT (Financial Times) Magazine*, August 2, 2013. http://www.ft.com/cms/s/2/4084c8ee-fa36-11e2-98e0-00144feabdc0.html (accessed September 4, 2015).

Cooper, J. Andrew G., and Orrin H. Pilkey. "Sea Level Rise and Shoreline Retreat: Time to Abandon the Bruun Rule." *Global and Planetary Change* 43 (2004): 157–71.

Correa, Ivan D., and Juan Louis Gonzalez. 2000. "Coastal Erosion and Village Relocation: A Colombian Case Study." *Ocean & Coastal Management* 43, no. 1 (2000): 51–64.

Dean, Cornelia. "Climate Change Harsh on Atlantic Barrier Islands." *PilotOnline*, September 30, 2014. http://hamptonroads.com/2014/09/climate-change-harsh-atlantic-barrier-islands (accessed September 4, 2015).

Dutton, A., A. E. Carlson, G. Milne, et al. "Sea-Level Rise Due to Polar Ice-Sheet Mass Loss During Past Warm Periods." *Science* 349, no. 6244 (2015). doi: 10.1126/science.aaa4019. (accessed August 20, 2015).

Ezer, Tal, Larry P. Atkinson, William B. Corlett, and Jose L. Blanco. "Gulf Stream's Induced Sea Level Rise and Variability Along the U.S. Mid-Atlantic Coast." *Journal of Geophysical Research: Oceans* 118 (2013): 685–97.

Fasullo, John T., Carmen Boening, Felix W. Landerer, and R. Steven Nerem. "Australia's Unique Influence of Global Sea Level in 2010–2011." *Geophysical Research Letters* 40, no. 16 (2013): 4368–73. http://dx.doi.org/10.1002/grl.50834; http://onlinelibrary.wiley.com/doi/10.1002/grl.50834/abstract (accessed September 4, 2015).

Fehrenbacher, Jill. "Inhabitat Interview: Koen Olthuis of Water-Studio.nl Talks About Design for a Water World." July 8, 2015. http://inhabitat.com/interview-koen-olthius-of-waterstudionl/ (accessed September 4, 2015).

Gardiner, Ned, and Emily Greenhalgh. "Viewing Sea Level Rise." October 30, 2013. National Oceanic and Atmospheric Administration. http://www.climate.gov/news-features/decision-makers-toolbox/viewing-sea-level-rise (accessed September 4, 2014).

Goddard, Paul B., Jianjun Yin, Stephen M. Griffies, and Shaoqing Zhang. "An Extreme Event of Sea-Level Rise Along the Northeast Coast of North America in 2009–2010." *Nature Communications* 6, no. 6346 (2014). doi: 10.1038/ncomms734. http://www.nature.com/ncomms/2015/150224/ncomms7346/pdf/ncomms7346.pdf (accessed September 4, 2015).

Goodell, Jeff. "Goodbye Miami." *Rolling Stone*, June 20, 2013, 94–103. http://www.rollingstone.com/politics/news/why-the-city-of-miami-is-doomed-to-drown-20130620 (accessed September 4, 2015).

Gornitz, Vivien. *Rising Seas: Past, Present, Future*. New York: Columbia University Press, 2013.

Greenfieldboyce, Nell. "What Would Happen If We Burned Up All of Earth's Fossil Fuels?" September 11, 2015. NPR. http://www.npr.org/sections/thetwo-way/2015/09/11/439538952/what-would-happen-if-we-burned-up-all-of-earths-fossil-fuels (accessed September 12, 2015).

Gregory, Kia. "Where Streets Flood with the Tide: A Debate over City Aid." *New York Times*, July 9, 2013. http://www.nytimes.com/2013/07/10/nyregion/debate-over-cost-and-practicality-of-protecting-part-of-queens-coast.html?pagewanted=1&_r=0&ref=kiagregory (accessed September 3, 2013).

Hansen, J., M. Sato, P. Hearty, et al. "Ice Melt, Sea Level Rise and Superstorms: Evidence from Paleoclimate Data, Climate Modeling, and Modern Observations That 2°C Global Warming Is Highly Dangerous." *Atmospheric Chemistry and Physics Discussions*, July 23, 2015, 20059–179. www.atmos-chem-phys-discuss.net/15/20059/2015/ doi:10.5194/acpd-15-20059-2015 (accessed August 30, 2015).

Hansen, James, et al. "Supplement of Ice Melt, Sea Level Rise and Superstorms: Evidence from Paleoclimate Data, Climate Modeling, and Modern Observations That 2°C Global Warming Is Highly Dangerous." *Atmospheric Chemistry and Physics Discussions*, July 23, 2015, 20059–179. doi: 10-5194/acpd-15-20059-2015-supplement (accessed August 31, 2015).

Hansen, James, Makiko Sato, Gary Russell, and Pushker Kharecha. "Climate Sensitivity, Sea Level and Atmospheric Carbon Dioxide." *Philosophical Transactions of the Royal Society A* 371 (2013): 20120294. http://dx.doi.org/10.1098/rsta.2012.0294 (accessed August 20, 2015).

Hertsgaard, Mark. "Climate Seer James Hansen Issues His Direst Forecast Yet." *Daily Beast*, July 20, 2015. http://www.thedailybeast.com/articles/2015/07/20/climate-seer-james-hansen-issues-his-direst-forecast-yet.html (accessed July 20, 2015).

Intergovernmental Panel on Climate Change. *Climate Change 2007: Synthesis Report*. www.ipcc.ch/pdf/assessment-report/ar4/syr/ar4_syr.pd (accessed August 11, 2015).

Intergovernmental Panel on Climate Change. *Climate Change 2013: The Physical Science Basis*. www.ipcc.ch/report/ar5/wg1 (accessed August 11, 2015).

Intergovernmental Panel on Climate Change. *Climate Change 2014: Mitigation of Climate Change*. www.ipcc.ch/report/ar5/wg3/ (accessed August 11, 2015).

Joyce, Christopher. "Climate Change Worsens Coastal Flooding from High Tides." October 8, 2014. NPR. http://www.npr.org/2014/10/08/354166982/climate-change-worsens-coastal-flooding-from-high-tides (accessed September 4, 2015).

Kahn, Brian, and Climate Central. "Sea Level Could Rise at Least 6 Meters." *Scientific American*, July 9, 2015. http://www.scientificamerican.com/article/sea-level-could-rise-at-least-6-meters/ (accessed July 12, 2015).

Khazendar, Ala, Christopher P. Borstad, Bernd Scheuchi, et al. "The Evolving Instability of the Remnant Larsen B Ice Shelf and Its Tributary Glaciers." *Earth and Planetary Letters* 419 (2015): 199–210. doi: 10.1016/j.epsl.2015.03.014 (accessed September 4, 2015).

Korten, Tristram. "In Florida, Officials Ban Term 'Climate Change.'" Florida Center for Investigative Reporting. *Newsletter*, March 8, 2015. fcir.org/2015/03/08/in-florida-officials-ban-term-climate-change/; *Miami Herald*. March 8, 2015. http://www.miamiherald.com/news/state/florida/article12983720.html (accessed August 11, 2015).

Levermann, Anders. "The Inevitability of Sea Level Rise." *Real Climate*, August 15, 2013. http://www.realclimate.org/index.php/archives/2013/08/the-inevitability-of-sea-level-rise/#more-15633 (accessed September 4, 2015).

Levermann, Anders, Peter U. Clark, Ben Marzeion, Glenn A. Milne, David Pollard, Valentina Radic, and Alexander Robinson. "The Multimillennial Sea-Level Commitment of Global Warming." *Proceedings of the National Academy of Sciences*, July 10, 2013. http://www.pnas.org/content/early/2013/07/10/1219414110 (accessed August 31, 2015).

McKie, Robin. "Miami The Great World City Is Drowning While the Powers That Be Look Away. *Guardian*, July 11, 2014. http://www.theguardian.com/world/2014/jul/11/miami-drowning-climate-change-deniers-sea-levels-rising (accessed September 4, 2015).

McNeill, Ryan, Deborah J. Nelson, and Duff Wilson. "As the Seas Rise, a Slow-Motion Disaster Gnaws at America's Shores." September 4, 2014. Thomsonreuters.com. http://www.reuters.com/investigates/special-report/waters-edge-the-crisis-of-rising-sea-levels (accessed August 12, 2015).

Mok, Kimberly. "Affordable Bamboo Housing Floats When It Floods." April 11, 2013. http://www.treehugger.com/green-architecture/affordable-disaster-resistant-bamboo-housing-floats-in-flood-hp-architects.html (accessed September 4, 2015).

Montaigne, Fen. "Leaving Our Descendants a Whopping Rise in Sea Levels: An Interview with Anders Levermann." *Yale Environment 360*, July 24, 2013. http://e360.yale.edu/feature/leaving_our_descendants_a_whopping_rise_in_sea_levels/2675/ (accessed September 4, 2015).

Mooney, Chris. "This Is What a Holy Shit Moment for Global Warming Looks Like." *Mother Jones*, May 12, 2014. www.motherjones.com/search/apachesoir_search/holy%20shit%62 (accessed August 11, 2015).

Mooney, Chris. "Why the Earth's Past Has Scientists So Worried About Sea Level Rise." *Washington Post*, July 9, 2015. http://www.washingtonpost.com/news/energy-environment/wp/2015/07/09/why-the-earths-past-has-scientists-so-worried-about-sea-level-rise/ (accessed September 2, 2015).

Neild, Barry. "Floating Snowflake Could Be the World's Coolest Hotel." August 5, 2014. CNN. http://www.cnn.com/2014/08/05/travel/snowflake-hotel/. Updated August 6, 2014 (accessed September 4, 2015).

O'Leary, Michael, Paul J. Hearty, William G. Thompson, Maureen E. Raymo, Jerry X. Milvovica, and Jody M. Webster. "Ice Sheet Collapse Following a Prolonged Period of Stable Sea Level During the Last Interglacial." *Nature Geoscience* 6, no. 9 (2013): 796–800. doi:10.1038/ngeo1890. www.nature.com/ngeo/journal/v6/n9/fall/ngeo1890.html (accessed September 4, 2015).

Olthuis, Koen, and David Keuning. "*Float: Building on Water to Combat Urban Congestion and Climate Change*." Amsterdam: Frame, 2010.

Paquette, Danielle. "Miami's Climate Catch-22: Building Waterfront Condos to Pay for Protection Against the Rising Sea." *Washington Post*, December 22, 2014. http://www.washingtonpost.com/news/storyline/wp/2014/12/22/miamis-climate-catch-22-building-luxury-condos-to-pay-for-protection-against-the-rising-sea/ (accessed September 4, 2015).

Parker, Laura. "Treading Water: Florida's Bill Is Coming Due as the Costs of Climate Change Add Up Around the Globe." *National Geographic*, February 2015. http://ngm.nationalgeographic.com/2015/02/climate-change-economics/parker-text (accessed September 4, 2015).

Pilkey, Orrin H., William J. Neal, Joseph T. Kelley, and J. Andrew G. Cooper. *The World's Beaches: A Global Guide to the Science of the Shoreline*. Berkeley: University of California Press, 2011.

Pilkey, Orrin H., and Keith C. Pilkey. *Global Climate Change: A Primer*. Durham, N.C.: Duke University Press, 2011.

Pilkey, Orrin H., and Rob Young. *The Rising Sea*. Washington, D.C.: Island Press, 2009.

Qiu, Jane. "Chain Reaction Shattered Huge Antarctica Ice Shelf." *Nature News*, August 9, 2013. http://www.nature.com/news/chain-reaction-shattered-huge-antarctica-ice-shelf-1.13540 (accessed September 4, 2015).

Rignot, Eric. "Global Warming: It's a Point of No Return in West Antarctica. What Happens Next?" *Guardian*, May 17, 2014. http://www.theguardian.com/commentisfree/2014/may/17/climate-change-antarctica-glaciers-melting-global-warming-nasa (accessed September 4, 2015).

Rohling, E. J., M. Fenton, F. J. Jorissen, et al. "Letters to Nature: Magnitudes of Sea-Level Lowstands of the Past 500,000 Years." *Nature* 394 (1998): 162–65. doi: 10.1038/28134 (accessed August 24, 2015).

Sallenger, Asbury H., Jr., Kara S. Doran, and Peter A. Howd. "Hotspot of Accelerated Sea-Level Rise on the Atlantic Coast of North America." *Nature Climate Change* 2 (2012): 884–88. Published online, June 24, 2012; corrected online, April 26, 2013. doi: 10.1038/nclimate1597 (accessed September 4, 2015)

Siddall, M., E. J. Rohling, A. Almogi-Labin, et al. "Sea-Level Fluctuations During the Last Glacial Cycle." *Nature* 423 (2003): 853–58. doi: 10.1038/nature01690 (accessed August 24, 2015).

Spanger-Siegfried, Erika, Melanie Fitzpatrick, and Kristina Dahl. "Encroaching Tides: How Sea Level Rise and Tidal Flooding Threaten U.S. East and Gulf Coast Communities over the Next 30 Years." October 2014. Union of Concerned Scientists. http://www.ucsusa.org/sites/default/files/attach/2014/10/encroaching-tides-full-report.pdf (accessed September 4, 2015).

Sweet, William, Joseph Park, John Marra, Chris Zervas, and Stephen Gill. "Sea Level Rise and Nuisance Flood Frequency Changes Around the United States." Technical Report NOS CO-OPS 073. Silver Spring, Md.: NOAA, June 2014.

Tessler, Z. D., C. J. Vörösmarty, M. Grossberg, et al. "Profiling Risk and Sustainability in Coastal Deltas of the World." *Science* 349, no. 6248 (2015): 638–43. doi: 10.1126/science.aab3574 (accessed August 28, 2015).

Viñas, Maria-Jose, and Carol Rasmussen. "Warming Seas and Melting Ice Sheets." August 26, 2015. NASA, Jet Propulsion Laboratory, and California Institute of Technology. http://www.jpl.nasa.gov/news/news.php?feature=4699 (accessed August 31, 2015).

Winkelmann, Ricarda, Anders Levermann, Andy Ridgwell, and Ken Caldeira. "Combustion of Available Fossil Fuel Resources Sufficient to Eliminate the Antarctic Ice Sheet." *Science Advances* 1, no. 8 (2015): e1500589. doi: 10.1126/sciadv.1500589.http://advances.sciencemag.org/content/1/8/e1500589.full (accessed September 12, 2015).

3. The Fate of Two Doomed Cities

Adams, Michael. "Florida's Citizens Under Scrutiny for Alleged Management Misconduct, Favoritism." *Insurance Journal*, November 23, 2012. http://www.insurancejournal.com/news/southeast/2012/11/23/271463.htm (accessed August 20, 2015).

Andersen, Christine F., Jurjen A. Battjes, David E. Daniel, Billy Edge, William Espey Jr., Robert B. Gilbert, Thomas L. Jackson, and David Kennedy. "The New Orleans Hurricane Protection System: What Went Wrong and Why: A Report by ASCE Hurricane Katrina External Review Panel." Reston, Va.: American Society of Civil Engineers, 2007. doi: http://dx.doi.org/10.1061/9780784408933. (accessed September 4, 2015).

"Application for Florida Reactors." *World Nuclear News*, July 23, 2009. http://www.world-nuclear-news.org/NN_Application_for_Florida_reactors_2307092.html (accessed July 22, 2010).

Berry, Leonard. "Impacts of Sea Level Rise on Florida's Domestic Energy and Water Infrastructure." Testimony before the U.S. Senate Committee on Energy and Natural Resources, April 19, 2012.

Blue Ribbon Committee. "Shoreline Management Final Report: Recommendations for Improved Beachfront Management in South Carolina." 2013. South Carolina Department of Health and Environmental Control. https://www.scdhec.gov/library/CR-010631.pdf (accessed September 4, 2015).

Brandt, Nadja. "Miami Building Boom Spreads into Downtown's Tent City." October 27, 2014. Bloomberg News. http://www.bloomberg.com/news/articles/2014-10-27/miami-building-boom-spreads-into-downtown-s-tent-city (accessed September 4, 2015).

Brannigan, Martha. "Miami's Condo Boom Redux." *Miami Herald*, July 6, 2014. http://www.miamiherald.com/news/business/biz-monday/article1974355.html (accessed August 20, 2015).

Brannigan, Martha, and Elaine Walker. "The Return of the Cranes: Miami-Dade Construction Projects on the Horizon in 2013." *Miami Herald*, January 3, 2013. http://www.miamiherald.com/2013/01/03/v-print/3166853/the-return-of-the-cranes-miami (accessed August 30, 2015).

Burkett, Virginia R., David B. Zilkoski, and David A. Hart. "Sea-Level Rise and Subsidence: Implications for Flooding in New Orleans, Louisiana." In *U.S. Geological Survey Subsidence Interest Group Conference: Proceedings of the Technical Meeting* (Galveston, Tex., November 27–29, 2001), edited by Keith R. Prince and Devin L. Galloway, pp. 63–70. 2003. Series no. 03-308. Washington, D.C.: U.S. Geological Survey, 2003. https://pubs.er.usgs.gov/publication/2000794 (accessed September 4, 2015).

City of New Orleans, Hazard Mitigation, Programs and Grants. "Hazard Mitigation Grant Program (HMGP)." 2010. http://www.nola.gov/hazard-mitigation/programs-and-grants/ (accessed September 4, 2015).

City of New Orleans, Office of Homeland Security Emergency Preparedness. "Orleans Parish, Louisiana: 2010 Hazard Mitigation Plan Update." March 7, 2011. http://www.nola.gov/getattachment/Hazard-Mitigation/Hazards-and-Planning/Orleans-Parish-2010-Hazard-Mitigation-Plan-Final-032311.pdf (accessed June 18, 2013).

Cunningham, Kevin J., Michael C. Sukop, Haibo Huang, Pedro F. Alvarez, H. Allen Curran, Robert A. Renken, and Joann F. Dixon. "Prominence of Ichnologically Influenced Macroporosity in the Karst Biscayne Aquifer: Stratiform "'Super-K'" Zones." *Geological Society of America Bulletin* 122, nos. 11–12 (2010): 2109. http://gsabulletin.gsapubs.org/content/122/11-12/2109.2 (accessed September 4, 2015).

de Diego, Javier. "South Miami Commissioners Propose Seceding from Florida." October 22, 2014. CNN. http://www.cnn.com/2014/10/22/us/south-florida-secession-new-state/ (accessed September 4, 2015).

El Akkad, Omar. "Come Hell or High Water: The Disaster Scenario That Is South Florida." *Globe and Mail*, July 17, 2015; last updated July 20, 2015. http://www.theglobeandmail.com/news/world/come-hell-or-high-water-the-disaster-scenario-that-is-south-florida/article25552300/ (accessed August 20, 2015).

Gardiner, Ned, and Emily Greenhalgh. "Viewing Sea Level Rise." October 30, 2013. National Oceanic and Atmospheric Administration. http://www.climate.gov/news-features/decision-makers-toolbox/viewing-sea-level-rise (accessed May 15, 2015.

Goldsmith, Wendi, Angela Desoto-Duncan, and Karen Durham-Aguilera. "Achieving the Unprecedented in New Orleans." *Military Engineer* 678 (2012): 50–53.

Goodell, Jeff. "Goodbye, Miami." *Rolling Stone*, July 4, 2013, 94–103. http://www.rollingstone.com/politics/news/why-the-city-of-miami-is-doomed-to-drown-20130620 (accessed September 4, 2015).

Institute for Sustainable Communities, Climate Leadership Academy. "Promising Practices in Adaptation & Resilience: A Resource Guide for Local Leaders, Version 1.0." Boston: Sustainable Communities Leadership Academy, September 10, 2010. http://sustainablecommunitiesleadershipacademy.org/resource_files/documents/Climate-Adaptation-Resource-Guide.pdf (accessed September 4, 2015).

Interagency Performance Evaluation Task Force, U.S. Army Corps of Engineers. "IPET Report on Flood Control During Katrina." June 2, 2006. http://www.flowcontrolnetwork.com/ipet-report-on-flood-control-during-katrina/ (accessed September 4, 2015).

Korten, Tristram. "In Florida, Officials Ban Term 'Climate Change.'" Florida Center for Investigative Reporting. *Newsletter*, March 8, 2015. fcir.org/2015/03/08/in-florida-officials-ban-term-climate-change/; *Miami Herald*. March 8, 2015. http://www.miamiherald.com/news/state/florida/article12983720.html (accessed August 11, 2015).

Marshall, Bob. "Experts: Talk Now About Drastic Changes, or Deal with Coastal Crisis Later." *The Lens*, September 15, 2015. http://thelensnola.org/2015/09/15/coastal-planners-talk-now-about-drastic-changes-or-deal-with-crisis-later/ (accessed September 27, 2015).

Marshall, Bob, Al Shaw, and Brian Jacobs. "Louisiana's Moon Shot." *The Lens / ProPublica, Knight-Mozilla OpenNews*, December 8, 2014. https://projects.propublica.org/larestoration/ (accessed December 10, 2014).

McKie, Robin. "Miami, the Great World City Is Drowning While the Powers That Be Look Away." *Guardian*, July 11, 2014. http://www.theguardian.com/world/2014/jul/11/miami-drowning-climate-change-deniers-sea-levels-rising (accessed September 4, 2015).

Meffert, Douglas J. "The Resilience of New Orleans: Urban and Coastal Adaptation to Disasters and Climate Change." Cambridge, Mass.: Lincoln Institute of Land Policy, August 2008. http://www.lincolninst.edu/pubs/1508_The-Resilience-of-New-Orleans; https://www.lincolninst.edu/pubs/dl/1508_744_Meffert%20Final.pdf (accessed September 4, 2015).

Meyer, Robert. "How Climate Change Is Fueling the Miami Real Estate Boom." *Business Week*, October 20, 2014. http://www.bloomberg.com/bw/articles/2014-10-20/how-climate-change-is-fueling-the-miami-real-estate-boom (accessed September 4, 2015).

National Weather Service. "Introduction to Storm Surge." No date. National Oceanic and Atmospheric Administration. www.nws.noaa.gov/om/hurricane/resources/surge_intro.pdf (accessed August 20, 2015).

Natural Resources Defense Council. "New Orleans, Louisiana: Identifying and Becoming More Resilient to Impacts of Climate Change." *NRDC Water Facts*, July 2011. http://www.nrdc.org/water/files/ClimateWaterFS_NewOrleansLA.pdf. (accessed September 4, 2015).

Nelson, Stephen A. "Why New Orleans Is Vulnerable to Hurricanes: Geologic and Historical Factors." Manuscript, Tulane University. Updated December 10, 2012. http://www.tulane.edu/~sanelson/New_Orleans_and_Hurricanes/New_Orleans_Vulnerability.htm (accessed June 14, 2013).

Nelson Institute for Environmental Studies. "Wetland Restoration and Community-Based Development Bayou Bienvenue, Lower Ninth Ward, New Orleans." Madison: University of Wisconsin: Water Resources Management Practicum, 2007, 2008. http://www.nelson.wisc.edu/docs/new_orleans_20081.pdf. Updated October 2009 (accessed July 11, 2013).

Nicholls, R. J., S. Hanson, Celine Herweijer, Nicola Patmore, Stéphane Hallegatte, Jan Corfee-Morlot, Jean Château, and Robert Muir-Wood. "Ranking Port Cities with High Exposure and Vulnerability to Climate Extremes: Exposure Estimates." OECD Environment Working Papers, no. 1. Paris: OECD, November 19, 2008. http://dx.doi.org/10.1787/011766488208 (accessed September 4, 2015).

Nuclear Regulatory Commission. "Turkey Point, Units 6 and 7 Application." July 2, 2009; updated February 25, 2014. Combined License Applications for New Reactors. http://www.nrc.gov/reactors/new-reactors/col/turkey-point.html (accessed September 4, 2015).

Paquette, Danielle. "Miami's Climate Catch-22: Building Waterfront Condos to Pay for Protection Against the Rising Sea." *Washington Post*, December 22, 2014. http://www.washingtonpost.com/news/storyline/wp/2014/12/22/miamis-climate-catch-22-building-luxury-condos-to-pay-for-protection-against-the-rising-sea/ (accessed September 4, 2015).

Parker, Laura. "Treading Water: Florida's Bill Is Coming Due as the Costs of Climate Change Add Up Around the Globe." *National Geographic*, February 2015. http://ngm.nationalgeographic.com/2015/02/climate-change-economics/parker-text (accessed September 4, 2015).

Prinos, S. T., M. A. Wacker, K. J. Cunningham, and D. V. Fitterman. "Origins and Delineation of Saltwater Intrusion in the Biscayne Aquifer and Changes in the Distribution of Saltwater in Miami-Dade County, Florida." U.S. Geological Survey Scientific Investigations Report 2014-5025. 2014. http://dx.doi.org/10.3133/sir20145025 (accessed September 4, 2015).

Sallenger, Asbury H., Jr., Kara S. Doran, and Peter A. Howd. "Hotspot of Accelerated Sea-Level Rise on the Atlantic Coast of North America." *Nature Climate Change* 2 (2012): 884–88. Published online, June 24, 2012; corrected online, April 26, 2013. doi: 10.1038/nclimate1597 (accessed September 4, 2015).

Schwartz, John. "Vast Defenses Now Shielding New Orleans." *New York Times*, June 14, 2012. http://www.nytimes.com/2012/06/15/us/vast-defenses-now-shielding-new-orleans.html?pagewanted=all&_r=0 (accessed September 4, 2015).

Spanger-Siegfried, Erika, Melanie Fitzpatrick, and Kristina Dahl. "Encroaching Tides: How Sea Level Rise and Tidal Flooding Threaten U.S. East and Gulf Coast Communities over the Next 30 Years." October 2014. Union of Concerned Scientists. http://www.ucsusa.org/sites/default/files/attach/2014/10/encroaching-tides-full-report.pdf (accessed September 4, 2015).

TetraTech. "Inner Harbor Navigation Canal Lake Borgne Surge Barrier." 2013. http://www.tetratech.com/en/projects/inner-harbor-navigation-canal-lake-borgne-surge-barrier (accessed August 31, 2015).

University of Wisconsin, Madison. "Welcome to the Bayou Bienvenue Wetland Triangle" (flyer). 2008. http://www.lakeforest.edu/live/files/1697-bayou-bienvenu-trianglepdf (accessed September 4, 2015).

U.S. Army Corps of Engineers. "Performance Evaluation of the New Orleans and Southeast Louisiana Hurricane Protection System: Draft Final Report of the Interagency Performance Evaluation Task Force." Vol. 1, "Executive Summary and Overview." June 1, 2006. http://www.nytimes.com/packages/pdf/national/20060601_ARMYCORPS_SUMM.pdf (accessed September 4, 2015).

U.S. Army Corps of Engineers. "Performance Evaluation of the New Orleans and Southeast Louisiana Hurricane Protection System: Final Report of the Interagency Performance Evaluation Task Force." January 29, 2007. http://www.iwr.usace.army.mil/Portals/70/docs/projects/29Jan07/IPET_Summary_USGS.pdf (accessed September 4, 2015).

U.S. Army Corps of Engineers, Team New Orleans. "Inner Harbor Navigation Canal Surge Barrier." 2011. http://www2.mvn.usace.army.mil/pd/projectslist/home.asp?projectID=300 (accessed June 18, 2013).

Wiedenman, Ryan. "Adaptive Response Planning for Sea Level Rise and Salt Water Intrusion in Miami Dade County." Master's thesis, Florida State University, 2010.

4. New and Old Amsterdam

Bagley, Katherine, and Marie Gallucci. "Is NYC's Climate Plan Enough to Win the Race Against Rising Seas?" *Inside Climate News*, June 20, 2013. http://insideclimatenews.org/news/20130620/nycs-climate-plan-enough-win-race-against-rising-seas (accessed September 4, 2015).

Beck, Graham T. "Plan to Expand Manhattan 500 Feet into the East River 'Financially Feasible' Says City." *Next City*, July 2, 2014. http://nextcity.org/daily/entry/plan-to-expand-manhattan-500-feet-into-the-east-river-financial-feasible (accessed September 4, 2015).

Centers for Disease Control and Prevention. "Deaths Associated with Hurricane Sandy—October–November 2012." *Morbidity and Mortality Weekly Report*, May 24, 2013. http://www.cdc.gov/mmwr/preview/mmwrhtml/mm6220a1.htm (accessed September 4, 2015).

Delta Commissie. "Working Together with Water." *Findings of the Deltacommissie*. 2008.

Diaz, Jason. "Klaus Jacob on the Future of a Post-Sandy New York City." *City Atlas* (blog), May 13, 2013. http://newyork.thecityatlas.org/lifestyle/klaus-jacob-future-post-sandy-york-city/ (accessed September 4, 2015).

Doyle, Alister, and Ryan McNeill. "Why Britain Is Flirting with Retreat from Its Battered Shores." December 12, 2014. Thomsonreuters.com. http://www.reuters.com/investigates/special-report/waters-edge-the-crisis-of-rising-sea-levels/#article-4-hard-choices (accessed September 4, 2015).

Goodyear, Sarah. "We're in This Together: What the Dutch Know About Flooding That We Don't." *Atlantic Citylab*, January 9, 2013. http://m.theatlanticcities.com/politics/2013/01/were-together-what-dutch-know-about-water-we-dont/4355/ (accessed September 4, 2015).

Gregory, Kia. "Where Streets Flood with the Tide: A Debate over City Aid." *New York Times*, July 9, 2013. http://www.nytimes.com/2013/07/10/nyregion/debate-over-cost-and-practicality-of-protecting-part-of-queens-coast.html?pagewanted=1&_r=0&ref=kiagregory (accessed September 3, 2013).

Harrison, Timothy. "After Hurricane Sandy and $250 Million in Property Buyouts, Memories, but Little Else, Will Remain in Oakwood Beach." June 5, 2014. SIlive.com. http://www.silive.com/specialreports/index.ssf/2014/06/after_hurricane_sandy_and_250.html. Updated June 5, 2014 (accessed September 4, 2015).

LaVorgna, Marc, and Lauren Passalacqua. "Mayor Bloomberg Presents the City's Long-Term Plan to Further Prepare for the Impacts of a Changing Climate." June 11, 2013. City of New York, Office of the Mayor. http://www1.nyc.gov/office-of-the-mayor/news/200-13/mayor-bloomberg-presents-city-s-long-term-plan-further-prepare-the-impacts-a-changing (accessed September 4, 2015).

Levermann, Anders, Peter U. Clark, Ben Marzeion, Glenn A. Milne, David Pollard, Valentina Radic, and Alexander Robinson. "The Multimillennial Sea-Level Commitment of Global Warming." *Proceedings of the National Academy of Sciences*, July 10, 2013. http://www.pnas.org/content/early/2013/07/10/1219414110 (accessed August 31, 2015).

McDermott, Maura. "New York Starts Buying Sandy-Damaged LI Homes." *Newsday*, June 5, 2014. http://www.newsday.com/classifieds/real-estate/new-york-starts-buying-sandy-damaged-li-homes-1.8350576 (accessed September 4, 2015).

New Jersey Department of Environmental Protection. "Frequently Asked Questions: Superstorm Sandy Blue Acres Buyout Program." http://www.nj.gov/dep/greenacres/pdf/faqs-blueacres.pdf. Updated April 29, 2015 (accessed September 4, 2015).

New York State, Governor's Office of Storm Recovery. "NY Rising Buyout and Acquisition Programs—Version 3.0." April 2015. https://stormrecovery.ny.gov/sites/default/files/uploads/po_20150415_buyout_and_acquisition_policy_manual_final_v3.pdf (accessed September 4, 2015).

New York State, Homes and Community Renewal, Office of Community Renewal. "State of New York Action Plan for Community Development Block Program Disaster Recovery." April 3 and 25, 2013. Federal Register Docket no. FR-5696-N-01. http://www.nyshcr.org/Publications/CDBGActionPlan.pdf (accessed September 4, 2015).

Province of Zuid-Holland. "The Sand Motor [Zandmotor: Delflandse Kust]." 2014. http://www.dezandmotor.nl/en-GB/the-sand-motor/ (accessed September 4, 2015).

Rijkswaterstaat, Province of South Holland. "The Sand Motor—Passionate Research." *YouTube* (video). April 17, 2014. http://www.youtube.com/watch?v=wtY4_QXcVsM&feature=youtu.be (accessed September 4, 2015).

Schechner, Sam. "Storm Barriers Likely to Close in New England." *Wall Street Journal*, August 27, 2011. http://www.wsj.com/news/articles/SB10001424053111904787404576534931624855042?mg=reno64-wsj&url=http%3A%2F%2Fonline.wsj.com%2Farticle%2FSB10001424053111904787404576534931624855042.html. (accessed September 4, 2015).

Scheurmann, Mathew. "Some on Staten Island Opt for Buyout of 'Houses That Don't Belong.'" September 19, 2014. NPR Cities Project. http://www.npr.org/2014/09/19/349559031/some-on-staten-island-opt-for-buyout-of-houses-that-dont-belong (accessed September 4, 2015).

5. Cities on the Brink

Bovarnick, Ben, Shiva Polefka, and Arpita Bhattacharyya. "Rising Waters, Rising Threat: How Climate Change Endangers America's Neglected Wastewater Infrastructure." October 2014. Center for American Progress. https://cdn.americanprogress.org/wp-content/uploads/2014/10/wastewater-report.pdf; www.infrastructureusa.org/rising-waters-rising-threat-how-climate-change-endangers-americas-neglected-wastewater-infrastructure/ (accessed September 5, 2015).

Brown, Sally, Abiy S. Kebede, and Robert J. Nichols. "Sea-Level Rise and Impacts in Africa, 2000 to 2100." School of Civil Engineering and the Environment, University of Southampton, revised April 11, 2011.

Englander, John. *High Tide on Main Street: Rising Sea Level and the Coming Coastal Crisis*. Science Bookshelf (online). 2012.

Goddard, Paul B., Jianjun Yin, Stephen M. Griffies, and Shaoqing Zhang. "An Extreme Event of Sea-Level Rise Along the Northeast Coast of North America in 2009–2010." *Nature Communications* 6, no. 6346 (2014). doi:10.1038/ncomms7346. http://www.nature.com/ncomms/2015/150224/ncomms7346/pdf/ncomms7346.pdf (accessed September 5, 2015).

Gornitz, Vivien. *Rising Seas: Past, Present, Future*. New York: Columbia University Press, 2013.

Hanson, Susan, Robert Nicholls, N. Ranger, S. Hallegatte, J. Corfee-Morlot, C. Herweijer, and J. Chateau. "A Global Ranking of Port Cities with High Exposure to Climate Extremes." *Climatic Change* 104 (2011): 89–111.

Idso, Craig, Fred Singer, et al. "Climate Change Reconsidered: 2009 Report of the Nongovernmental International Panel on Climate Change (NIPCC)." Chicago: Heartland Institute, 2009. https://www.heartland.org/sites/default/files/NIPCC%20Final.pdf (accessed August 31, 2015).

Kelley, Joseph T., Orrin H. Pilkey, and J. Andrew G. Cooper, eds. *America's Most Vulnerable Coastal Communities*. Geological Society of America Special Paper, no. 460. 2009.

Lakshmi, Ahana, Aurofilio Schiavina, Probir Banerjee, Agit Reddy, Sunaina Mandeen, Sudarshan Rodriguez, and Deepak Apte. "The Challenged Coast of India: A Report." 2012. Prepared by PondyCAN in collaboration with BNHS and TISS. Released on the occasion of COP11 (Convention on Biological Diversity), Hyderabad, India. October 8–19, 2012. http://deepakapte.com/attachments/article/20/Challenged%20Coast%20of%20India_Lowres.pdf (accessed September 5, 2015).

McNeill, Ryan, Deborah J. Nelson, and Duff Wilson. "As the Seas Rise, a Slow-Motion Disaster Gnaws at America's Shores." September 4, 2014. Thomsonreuters.com. http://www.reuters.com/investigates/special-report/waters-edge-the-crisis-of-rising-sea-levels (accessed August 12, 2015).

Nelson, Deborah J., Ryan McNeill, and Duff Wilson. "Why Americans Are Flocking to Their Sinking Shore Even as the Risks Mount." September 17, 2014. Thomsonreuters.com. http://www.reuters.com/investigates/special-report/waters-edge-the-crisis-of-rising-sea-levels/ (accessed August 12, 2015).

Nelson, Deborah J., and Duff Wilson. 2014. "Where Retreating—Now or Later—Is the Only Option." September 2014. Thomsonreuters.com. http://www.reuters.com/investigates/special-report/waters-edge-the-crisis-of-rising-sea-levels/ (accessed August 12, 2015).

Nicholls, R. J., S. Hanson, Celine Herweijer, Nicola Patmore, Stéphane Hallegatte, Jan Corfee-Morlot, Jean Château, and Robert Muir-Wood. "Ranking Port Cities with High Exposure and Vulnerability to Climate Extremes: Exposure Estimates." OECD Environment Working Papers, no. 1. Paris: OECD, November 19, 2008. http://dx.doi.org/10.1787/011766488208 (accessed September 5, 2015).

O'Malley, Nick. "Warming World's Rising Seas Wash Away Some of South Florida's Glitz." *Sydney Morning Herald*, December 20, 2014. http://www.smh.com.au/world/warming-worlds-rising-seas-wash-away-some-of-south-floridas-glitz-20141218-129wub.html (accessed September 5, 2015).

Radford, Tim. "U.S. Coastal Cities Face Daily Flooding by Mid-Century—NOAA." *RTCC News*, January 6, 2015. http://www.rtcc.org/2015/01/06/us-coastal-cities-face-daily-flooding-by-mid-century-noaa/ (accessed September 6, 2015).

Seasholes, Nancy S. *Gaining Ground: A History of Landmaking in Boston*. Cambridge, Mass.: MIT Press, 2003.

Sheppard, Kate. "Flood, Rebuild, Repeat: Are We Ready for a Superstorm Sandy Every Other Year?" *Mother Jones*, July–August 2013. http://www.motherjones.com/environment/2013/07/hurricane-sandy-global-warming-flooding (accessed November 29, 2014).

Spanger-Siegfried, Erika, Melanie Fitzpatrick, and Kristina Dahl. "Encroaching Tides: How Sea Level Rise and Tidal Flooding Threaten U.S. East and Gulf Coast Communities over the Next 30 Years." October 2014. Union of Concerned Scientists. http://www.ucsusa.org/global_warming/impacts/effects-of-tidal-flooding-and-sea-level-rise-east-coast-gulf-of-mexico#.VEBh2ChELAN (accessed September 6, 2015).

Tarrant, Bill. "In Jakarta, That Sinking Feeling Is All Too Real." December 22, 2014. Thomsonreuters.com. http://www.reuters.com/article/2014/12/22/us-sealevel-subsidence-jakarta-sr-idUSKBN0K016S20141222 (accessed September 6, 2015).

United Nations. "World Urbanization Prospects." 2014. http://www.un.org/en/development/desa/publications/2014-revision-world-urbanization-prospects.html (accessed September 6, 2015).

United Nations, Department of Economic and Social Affairs, Population Division. "World Population Prospects: The 2012 Revision." 2013. http://esa.un.org/wpp/Documentation/publications.htm (accessed September 6, 2015).

World Health Organization. "Report from the Intergovernmental Panel on Climate Change (IPCC): Climate Change 2014: Impacts, Adaptation, and Vulnerability." 2014. http://www.who.int/globalchange/environment/climatechange-2014-report/en/ (accessed August 13, 2015).

6. The Taxpayers and the Beach House

Bauerlein, Valerie. "How to Measure a Storm's Fury One Breakfast at a Time: Disaster Pros Look to 'Waffle House Index'; State of the Menu Gives Clue to Damage." *Wall Street Journal*, September 1, 2011. http://www.wsj.com/articles/SB10001424053111904716604576542460736605364 (accessed September 9, 2015).

"Citizens Property Insurance Corporation." *Wikipedia*. Last modified February 14, 2015. http://en.wikipedia.org/wiki/Citizens_Property_Insurance_Corporation; https://www.citizensfla.com/ (accessed September 6, 2015).

Cleetus, Rachel. "Overwhelming Risk: Rethinking Flood Insurance in a World of Rising Seas." August 2013; revised February 2014. Union of Concerned Scientists. www.ucsusa.org/floodinsurance (accessed September 6, 2015).

Coburn, Tom. "Washed Out to Sea: How Congress Prioritizes Beach Pork over National Needs." May 2009. U.S. Senate, 111th Cong. Congressional Oversight & Investigation Report. Office of Senator Tom Coburn, M.D. https://www.hsdl.org/?view&did=726063 (accessed September 4, 2015).

Daley, Beth. "Oceans of Trouble for U.S. Taxpayers." March 9, 2014. New England Center for Investigative Reporting. http://necir.org/2014/03/09/oceans-of-trouble-for-u-s-taxpayers/ (accessed September 6, 2015).

Federal Emergency Management Agency. "Coastal Barrier Resources System." http://www.fema.gov/coastal-barrier-resources-system. Last updated April 26, 2015 (accessed September 6, 2015).

Federal Emergency Management Agency. *FEMA Property Acquisition Handbook for Local Communities*. Washington, D.C.: FEMA, 2014. https://www.fema.gov/media-library/assets/documents/3117. Last updated May 1, 2014 (accessed September 6, 2015).

Federal Emergency Management Agency. "Hazard Mitigation Grant Program." http://www.fema.gov/hazard-mitigation-grant-program. Last updated August 25, 2015 (accessed September 6, 2015).

Federal Emergency Management Agency. "Pre-Disaster Mitigation Grant Program." http://www.fema.gov/pre-disaster-mitigation-grant-program. Last updated August 25, 2015 (accessed September 6, 2015).

Fernandes, Deirdre. "Changes to Flood Insurance Mean Higher Costs." *Boston Globe*, October 16, 2013. http://www.bostonglobe.com/business/2013/10/15/changes-flood-insurance-mean-higher-costs-water/1hXoKECepAhcpothIhnfeI/story.html (accessed September 6, 2015).

Gillis, Justin, and Felicity Barringer. "As Coasts Rebuild and U.S. Pays, Repeatedly, the Critics Ask Why." *New York Times*, November 18, 2012. http://www.nytimes.com/2012/11/19/science/earth/as-coasts-rebuild-and-us-pays-again-critics-stop-to-ask-why.html (accessed September 6, 2015).

Gornitz, Vivien. *Rising Seas: Past, Present, Future*. New York: Columbia University Press, 2013.

Haddow, George, Jane A. Bullock, and Kim Haddow. *Global Warming, Natural Hazards, and Emergency Management*. Boca Raton, Fla.: CRC Press, 2008.

Harkinson, Josh. "Beyoncé's Beach House Brouhaha." *Mother Jones*, April 27, 2010. http://m.motherjones.com/blue-marble/2010/04/beyonc%C3%A9s-beach-house-brouhaha (accessed September 6, 2015).

Harkinson, Josh. "Buh-bye East Coast Beaches." *Mother Jones*, April 27, 2010. http://www.motherjones.com/environment/2010/04/climate-desk-sea-level-rise-epa?page=2 (accessed September 6, 2015).

Lehmann, R. J. "R Street Warns Congress Not to Gut Crucial Flood Insurance Reforms." Washington, D.C.: R Street Institute, October 28, 2013. http://www.rstreet.org/news-release/r-street-warns-congress-not-to-gut-crucial-flood-insurance-reforms/ (accessed September 6, 2015).

McNeill, Ryan, Deborah J. Nelson, and Duff Wilson. "As the Seas Rise, a Slow-Motion Disaster Gnaws at America's Shores." September 4, 2014. Thomsonreuters.com. http://www.reuters.com/investigates/special-report/waters-edge-the-crisis-of-rising-sea-levels (accessed August 12, 2015).

Meyer, Theodoric. "Four Ways the Government Subsidizes Risky Coastal Building." June 19, 2013. ProPublica. http://www.propublica.org/article/four-ways-the-government-subsidizes-risky-coastal-rebuilding (accessed September 6, 2015).

Peterson, Bo. "Capt. Sam's Spit, New Storm Wave Device Also Part of S.C. House 'Seawall' Bill." *Post and Courier*, May 11, 2014. http://www.postandcourier.com/article/20140511/PC16/140519929 (accessed September 6, 2015).

Pierce, Walter. "Citizens United (in Debt)." *INDreporter*, November 9, 2012. http://www.theind.com/news/indreporter/11946-citizens-united-in-debt (accessed September 6, 2015).

Plumer, Brad. "Congress Tried to Cut Subsidies for Homes in Flood Zones. It Was Harder Than They Thought." *Wonkblog*, *Washington Post*, October 30, 2013. http://www.washingtonpost.com/blogs/wonkblog/wp/2013/10/30/congress-tried-to-stop-subsidizing-homes-in-flood-zones-it-was-harder-than-they-thought/ (accessed September 6, 2015).

Rocky Mountain Insurance Information Association. "Catastrophe Facts & Statistics." 2014. http://www.rmiia.org/Catastrophes_and_Statistics/catastrophes.asp (accessed September 6, 2015).

Sheppard, Kate. "Flood Grant Program to Let Communities Consider Rising Seas Due to Climate Change." *Huffington Post*, December 23,

2013. http://www.huffingtonpost.com/2013/12/23/fema-flood-program_n_4495117.html;http://www.floodsmart.gov/floodsmart/pages/about/nfip_overview.jsp (accessed September 6, 2015).

Siders, Anne. "Managed Coastal Retreat: A Legal Handbook on Shifting Development Away from Vulnerable Areas." October 2013. Columbia Center for Climate Change Law, Columbia Law School. http://web.law.columbia.edu/sites/default/files/microsites/climate-change/files/Publications/Fellows/ManagedCoastalRetreat_FINAL_Oct%2030.pdf (accessed September 6, 2015).

State of Texas Constitution and Statutes. "Natural Resources Code, Title 2, Subtitle E, Chap. 61: Use and Maintenance of Public Beaches." Statutes on this website are current through August 2013. http://www.statutes.legis.state.tx.us/Docs/NR/htm/NR.61.htm (accessed September 6, 2015).

Stossel, John. "Give Me a Break." September 20, 2014. ABC News. http://abcnews.go.com/2020/GiveMeABreak/story?id=123653&page=1&singlePage=true (accessed September 6, 2015).

U.S. Department of Housing and Urban Development. Docket no. FR-5696-N-06. Second Allocation, Waivers, and Alternative Requirements for Grantees Receiving Community Development Block Grant (CDBG) Disaster Recovery Funds in Response to Hurricane Sandy. Applicable Rules, Statutes, Waivers, and Alternative Requirements. "Comprehensive Risk Analysis." *Federal Register* 78, no. 222. November 18, 2013/Notices. 78 FR 69107, §VI(2)(d).

Veneziani, Vince. " Beyoncé Knowles Gets $425,000 Taxpayer Bailout on Her House in Texas." *Business Insider*, April 27, 2010. http://www.businessinsider.com/beyonce-knowles-gets-425000-taxpayer-bailout-on-her-house-in-texas-2010-4?op=1 (accessed September 6, 2015).

Yu, Elly. "When Disaster Strikes, FEMA Turns to Waffle House." *Marketplace*, March 4, 2015. http://www.marketplace.org/topics/business/when-disaster-strikes-fema-turns-waffle-house (accessed September 6, 2015).

Zilbermints, Regina. "FEMA's Fugate Explains 'Waffle House index,' Tours School Safe Rooms During South Mississippi Visit." *Sun Herald*, August 26, 2015. http://www.sunherald.com/2015/08/26/6383801_fema-mema-directors-tour-school.html?rh=1 (accessed September 6, 2015).

7. Coastal Calamities

Bender, Hannah. "10 States at Greatest Risk of Storm Surge Damage." *Property Casualty 360°*, June 6, 2013. http://www.propertycasualty360.com/2013/06/06/10-states-at-greatest-risk-of-storm-surge-damage (accessed September 6, 2015).

Botts, Howard, Thomas Jeffery, Wei Du, and Logan Suhr. "2014 CoreLogic Storm Surge Report." July 2014. http://www.corelogic.com/research/storm-surge/corelogic-2014-storm-surge-report.pdf (accessed September 6, 2015).

Botts, Howard, Wei Du, Thomas Jeffery, Steven Kolk, Zachary Pennycook, and Logan Surh. "Corelogic Storm Surge Report." 2013. http://www.corelogic.com/research/storm-surge/corelogic-2013-storm-surge-report.pdf (accessed September 6, 2015).

Cronin, William B. *The Disappearing Islands of the Chesapeake*. Baltimore: Johns Hopkins University Press, 2005.

Davis, Tony. "Underwater Cities: Climate Change Begins to Reshape the Urban Landscape." October 27, 2011. Grist.org. http://grist.org/cities/2011-10-26-underwater-cities-climate-change-begins-reshape-urban-landscape/ (accessed September 6, 2015).

Funk, McKenzie. *Windfall: The Booming Business of Global Warming*. New York: Penguin Press, 2014.

Gillis, Justin. "The Flood Next Time." *New York Times*, January 13, 2014. http://www.nytimes.com/2014/01/14/science/earth/grappling-with-sea-level-rise-sooner-not-later.html?smid=tw-share&_r=1 (accessed September 6, 2015).

Goldberg, Jeffrey. ""Drowning Kiribati." *Bloomberg Businessweek*, November 21, 2013. www.bloomberg.com/bw/articles/2013-11-21/kiribati-climate-change-destroys-pacific-island-nation (accessed August 13, 2015).

Gornitz, Vivien. *Rising Seas: Past, Present, Future*. New York: Columbia University Press, 2013.

Griggs, Gary B. "Lost Neighborhoods of the California Coast." *Journal of Coastal Research On-line*, November 18, 2013. http://dx.doi.org/10.2112/13A-00007.1 (accessed September 4, 2015).

Hanscom, Greg. "Sea Swallows the Last House in Doomed Virginia Beach Town." December 9, 2014. Grist.org. http://grist.org/climate-energy/sea-swallows-the-last-house-in-doomed-virginia-beach-town/ (accessed January 20, 2015).

Heberger, Matthew, Heather Cooley, Pablo Herrera, Peter Gleick, and Eli Moore. "The Impacts of Sea Level Rise on the California Coast." Pacific Institute, California Climate Change Center, 2009.

Intergovernmental Panel on Climate Change. "Climate Change 2014: Impacts, Adaptation, and Vulnerability." 2014. http://www.who.int/globalchange/environment/climatechange-2014-report/en/ (accessed August 13, 2015).

McNeill, Ryan, Deborah J. Nelson, and Duff Wilson. "As the Seas Rise, a Slow-Motion Disaster Gnaws at America's Shores." September 4, 2014. Thomsonreuters.com. http://www.reuters.com/investigates/special-report/waters-edge-the-crisis-of-rising-sea-levels (accessed August 12, 2015).

Montgomery, David. "Crisfield, Md., Beats Back a Rising Chesapeake Bay." *Washington Post* (magazine), October 24, 2013. https://www.washingtonpost.com/lifestyle/magazine/crisfield-md-beats-back-a-rising-chesapeake-bay/2013/10/24/ab213bda-0f1f-11e3-85b6-d27422650fd5_story.html (accessed September 6, 2015).

National Hurricane Center. "Storm Surge Overview." Miami: National Weather Service, 2013. http://www.nhc.noaa.gov/surge/. Last modified September 5, 2014 (accessed September 6, 2015).

Pilkey, Orrin H., and Mary Edna Fraser. *A Celebration of the World's Barrier Islands*. New York: Columbia University Press, 2003.

Pilkey, Orrin H., William J. Neal, Joseph T. Kelley, and J. Andrew G. Cooper. *The World's Beaches: A Global Guide to the Science of the Shoreline*. Berkeley: University of California Press, 2011.

Pilkey, Orrin H., and Rob Young. *The Rising Sea*. Washington, D.C.: Island Press, 2009.

Wilkinson, Peter. "Sea Level Rise Could Cost Port Cities $28 Trillion." November 23, 2009. CNN Tech: Going Green. http://www.cnn.com/2009/TECH/science/11/23/climate.report.wwf.allianz/ (accessed September 6, 2015).

Wilson, Duff. "The Village That Must Move—But Can't." 2014. *Thomsonreuters.com*. http://www.reuters.com/investigates/special-report/waters-edge-the-crisis-of-rising-sea-levels/#article-3-grand-designs (accessed September 6, 2015).

8. Drowning in Place

"Application for Florida Reactors." *World Nuclear News*, July 23, 2009. http://www.world-nuclear-news.org/NN_Application_for_Florida_reactors_2307092.html (accessed July 22, 2010).

Berry, Leonard. "Impacts of Sea Level Rise on Florida's Domestic Energy and Water Infrastructure." Testimony before the U.S. Senate Committee on Energy and Natural Resources. April 19, 2012.

Bovarnick, Ben, Shiva Polefka, and Arpita Bhattacharyya. "Rising Waters, Rising Threat: How Climate Change Endangers America's Neglected Wastewater Infrastructure." October 2014. Center for American Progress. https://cdn.americanprogress.org/wp-content/uploads/2014/10/wastewater-report.pdf; www.infrastructureusa.org/rising-waters-rising-threat-how-climate-change-endangers-americas-neglected-wastewater-infrastructure/ (accessed September 5, 2015).

Bullock, Jane A., George D. Haddow, and Kim S. Haddow, eds. *Global Warming, Natural Hazards, and Emergency Management*. Boca Raton, Fla.: CRC Press, 2009.

Cardno ENTRIX. *Joint Permit Application Package: Delaware City Refining Company, Shoreline Stabilization & Restoration Project*. New Castle, Del.: Cardno Entrix, May 2014. https://delaware.sierraclub.org/sites/delaware.sierraclub.org/files/documents/2014/07/Joint%20Permit%20Application%20Cover%20and%20TOC.pdf (accessed September 12, 2015).

Chacko, Sarah. "Bases at Risk." 2014. Global Warning: National Security Journalism Initiative. http://global-warning.org/main/installations/ (accessed September 6, 2015).

Daggett, Stephen. "Quadrennial Defense Review 2010: Overview and Implications for National Security Planning." May 17, 2010. https://www.fas.org/sgp/crs/natsec/R41250.pdf (accessed July 29, 2015).

Flynn, Timothy J., Stuart G. Walesh, James G. Titus, and Michael C. Barth. "Implications of Sea Level Rise for Hazardous Waste Sites in Coastal Floodplains." In *Greenhouse Effect and Sea Level Rise: A Challenge for This Generation*, edited by Michael C. Barth and James G. Titus, 271–94. New York: Van Nostrand Reinhold, 1984. http://www.papers.risingsea.net/downloads/Challenge_for_this_generation_Barth_and_Titus_chapter9.pdf (accessed September 6, 2015).

Guo, Jeff. "The Old Man and the Rising Sea." *Washington Post*, December 2, 2014. http://www.washingtonpost.com/news/storyline/wp/2014/12/02/the-old-man-and-the-rising-sea-2/ (accessed December 26, 2014).

Holtz, Debra, Adam Markham, Kate Cell, and Brenda Ekwurzel. "National Landmarks at Risk: How Rising Seas, Floods, and Wildfires Are Threatening the United States' Most Cherished Historic Sites." May 2014. Union of Concerned Scientists. http://www.ucsusa.org/sites/default/files/legacy/assets/documents/global_warming/National-Landmarks-at-Risk-Full-Report.pdf (accessed September 6, 2015).

Jaffe, Eric. "Why Sewage Plants Are Especially Vulnerable to Climate Change." *Atlantic CityLab*, May 2, 2013. http://www.theatlanticcities.com/jobs-and-economy/2013/05/why-sewage-plants-are-especially-vulnerable-climate-change/5464/ (accessed September 6, 2015).

Kenward, Alyson, Daniel Yawitz, and Urooj Raja. "Sewage Overflows from Hurricane Sandy." April 30, 2013. Climate Central. http://www.climatecentral.org/pdfs/Sewage.pdf; www.climatecentral.org/news/11-billion-gallons-of-sewage-overflow-from-hurricane-sandy-15924 (accessed September 6, 2015).

McNeill, Ryan, Deborah J. Nelson, and Duff Wilson. "As the Seas Rise, a Slow-Motion Disaster Gnaws at America's Shores." September 4, 2014. Thomsonreuters.com. http://www.reuters.com/investigates/special-report/waters-edge-the-crisis-of-rising-sea-levels. (accessed August 12, 2015).

National Oceanic and Atmospheric Administration. "State of the Coast." http://stateofthecoast.noaa.gov/; stateofthecoast.noaa.gov/features/reports.html. Revised March 25, 2013 (accessed September 6, 2015).

Nuclear Regulatory Commission. "Turkey Point, Units 6 and 7 Application." Combined License Applications for New Reactors. July 2, 2009; updated February 25, 2014. http://www.nrc.gov/reactors/new-reactors/col/turkey-point.html (accessed September 4, 2015).

Shifflett, Shane, and Kate Sheppard. "How Rising Seas Could Sink Nuclear Plants on the East Coast." *Huffington Post*, May 19, 2014. www.huffingtonpost.com/2014/05/19/maps-rising-seas-storms-threaten-flood-coastal-nuclear-power-plants_n_5233306.html (accessed September 6, 2015).

Sierra Club. "Plan for Refinery Shoreline Stabilization Project Raises Questions: Sierra Club Request Public Hearing." July 22, 2014. Sierra Club, Delaware Chapter. http://delaware.sierraclub.org/content/plan-refinery-shoreline-stabilization-project-raises-questions-sierra-club-requests-public (accessed September 12, 2015).

Strauss, Ben, and Remik Ziemlinski. "Sea Level Rise Threats to Energy Infrastructure: A Surging Seas Brief Report." April 19, 2012. Climate Central. http://slr.s3.amazonaws.com/SLR-Threats-to-Energy-Infrastructure.pdf (accessed September 6, 2015).

U.S. Energy Information Administration. "Number and Capacity of Petroleum Refineries." June 19, 2015. www.eia.gov/dnav/pet/pet_pnp_cap1_dcu_nus_a.htm (accessed September 9, 2015).

9. The Cruelest Wave

Amnesty International. "Amnesty International Releases Report on Human Rights Violations in Gulf Coast Recovery Efforts." Press release, April 9, 2010. http://www.amnestyusa.org/news/press-releases/usa-amnesty-international-releases-report-on-human-rights-violations-in-gulf-coast-recovery-efforts (accessed September 6, 2015).

Ayre, James. "Tesla CEO Elon Musk: Current Refugee Crisis Just Small Taste of What Climate Change Could Bring." September 29, 2015. CleanTechnica. http://cleantechnica.com/2015/09/29/tesla-ceo-elon-musk-current-refugee-crisis-just-small-taste-climate-change-bring/ (accessed October 2, 2015).

Banerjee, Bidisha. "The Great Wall of India." *Slate*, December 20, 2010. http://www.slate.com/articles/health_and_science/green_room/2010/12/the_great_wall_of_india.html (accessed September 6, 2015).

Brown, Donald A. *Climate Change Ethics: Navigating the Perfect Moral Storm*. London: Routledge, 2013.

Brown, Oli. "Migration and Climate Change." International Organization for Migration Research Series, no. 31. Geneva: IOM, 2008. http://www.iom.cz/files/Migration_and_Climate_Change_-_IOM_Migration_Research_Series_No_31.pdf (accessed July 30, 2015).

Brown, Sally, Abiy S. Kebede, and Robert J. Nicholls. "Sea-Level Rise and Impacts in Africa, 2000 to 2100." Revised April 11, 2011. United

Nations Environment Programme. Climate Change. http://www.unep.org/climatechange/adaptation/Portals/133/documents/AdaptCost/9%20Sea%20Level%20Rise%20Report%20Jan%202010.pdf (accessed July 2014).

Chameides, Bill. "The Carterets' Big Move: A Sign of the Future?" *Green Grok* (blog), March 31, 2009. Nicholas School of the Environment, Duke University. http://blogs.nicholas.duke.edu/thegreengrok/carteretislands/ (accessed September 6, 2015).

Egan, Timothy. *The Worst Hard Time: The Untold Story of Those Who Survived the Great American Dust Bowl*. Boston: Houghton Mifflin, 2006.

Friedman, Lisa. "A Global 'National Security' Issue Lurks at Bangladesh's Border." *New York Times*, March 23, 2009. www.nytimes.com/cwire/2009/03/23/23climatewire-a-global-national-security-issue-lurks-at-ba-10247.html?pagewanted=all (accessed September 6, 2015).

Gardiner, Stephen M. *A Perfect Moral Storm: The Ethical Tragedy of Climate Change*. Oxford: Oxford University Press, 2011.

Goldberg, Jeffrey. "Drowning Kiribati." *Bloomberg Businessweek*, November 21, 2013. www.bloomberg.com/bw/articles/2013-11-21/kiribati-climate-change-destroys-pacific-island-nation (accessed August 13, 2015).

Goldberg, Ruby. "India's 'Fence of Shame.'" *New Yorker*, December 18, 2013. http://www.newyorker.com/culture/photo-booth/indias-fence-of-shame (accessed September 6, 2015).

Gornitz, Vivien. *Rising Seas: Past, Present, Future*. New York: Columbia University Press, 2013.

Government Accountability Office. "Alaska Native Villages: Limited Progress Has Been Made on Relocating Villages Threatened by Flooding and Erosion." Report GAO 09-551 to Congressional Requesters. June 3, 2009. http://www.gao.gov/assets/300/290468.pdf; http://www.gao.gov/products/GAO-09–551 (accessed September 4, 2015).

Hayes, Chris. "The Climate Debate: Director James Cameron on Efforts to Fight Climate Change." *All in with Chris Hayes*, MSNBC (video), September 15, 2015. http://www.msnbc.com/all-in/watch/james-cameron-on-fighting-climate-change-526231619971 (accessed September 15, 2015).

Intergovernmental Panel on Climate Change. "Climate Change 2014: Impacts, Adaptation, and Vulnerability." 2014. http://www.who.int/globalchange/environment/climatechange-2014-report/en/ (accessed August 13, 2015).

Kauffman, Alexander C. "Elon Musk Says Climate Change Refugees Will Dwarf Current Crisis." *Huffington Post*, September 24, 2015. http://www.huffingtonpost.com/entry/elon-musk-in-berlin_560484dee4b08820d91c5f5f (accessed October 2, 2015).

Keane, Sandi. "Kiribati and the Coming Climate Refugee Crisis." *Independent Australia*, November 29, 2013. http://www.independentaustralia.net/article-display/kiribati-and-the-coming-climate-refugee-crisis,5935 (accessed September 6, 2015).

Kelley, Colin, Shahrzad Mohtadi, Mark A. Cane, et al. "Climate Change in the Fertile Crescent and Implications of the Recent Syrian Drought." *Proceedings of the National Academy of Sciences of the United States* 112, no. 11 (2015): 3241–46. http://www.pnas.org/content/112/11/3241.abstract. doi: 10.1073/pnas.1421533112 (accessed September 16, 2015).

Lovgren, Stefan. "Climate Change Creating Millions of 'Eco Refugees,' UN Warns." *National Geographic News*, November 18, 2005. http://news.nationalgeographic.com/news/2005/11/1118_051118_disaster_refugee.html (accessed September 6, 2015).

Mydans, Seth. "Vietnam Finds Itself Vulnerable If Sea Rises." *New York Times*, September 23, 2009. www.nytimes.com/2009/09/24/world/asia/24delta.html?pagewanted=all&_r=0 (accessed August 13, 2015).

Myers, Norman. "Environmental Refugees, an Emergent Security Issue." Paper presented at the Thirteenth Economic Forum, May 23–27, 2005, Prague, Czech Republic. http://www.osce.org/eea/14851?download=true. (accessed September 4, 2015). Based on Myers, N., and J. Kent. *Environmental Exodus: An Emergent Crisis in the Global Arena*. Washington, D.C.: Climate Institute; and Myers, N. "Environmental Refugees: Our Latest Understanding." *Philosophical Transactions of the Royal Society B* 356 (1995): 16.1–16.5.

Nicholls, R. J., S. Hanson, Celine Herweijer, Nicola Patmore, Stéphane Hallegatte, Jan Corfee-Morlot, Jean Château, and Robert Muir-Wood. "Ranking Port Cities with High Exposure and Vulnerability to Climate Extremes: Exposure Estimates." OECD Environment Working Papers, no. 1. Paris: OECD, November 19,

2008. http://dx.doi.org/10.1787/011766488208 (accessed September 5, 2015).

Quigley, J. T. "New Zealand 'Climate Change Refugee' Rejected by High Court." *Diplomat*, November 27, 2013. http://thediplomat.com/2013/11/new-zealand-climate-change-refugee-rejected-by-high-court/ (accessed September 6, 2015).

Seabrook, John. "The Beach Builders: Can the Jersey Shore Be Saved?" *New Yorker*, July 22, 2013. http://www.newyorker.com/magazine/2013/07/22/the-beach-builders (accessed September 6, 2015)).

Smith, Tierney. "Pacific Islander Loses Climate Refugee Bid in New Zealand." November 28, 2013. TckTckTck: The Global Call for Climate Action. http://tcktcktck.org/2013/11/pacific-islander-looses-climate-refugee-bid-new-zealand/59042 (accessed September 6, 2015).

Stenhouse, Neil, Edward Maibach, Sara Cobb, et al. "Meteorologists' Views About Global Warming: A Survey of American Meteorological Society Professional Members." *Bulletin of the American Meteorological Society*, July 2014. http://journals.ametsoc.org/doi/pdf/10.1175/BAMS-D-13–00091.1 (accessed August 13, 2015).

Welch, Craig. "Climate Change Helped Spark Syrian War, Study Says: Research Provides First Deep Look at How Global Warming May Already Influence Armed Conflict." *National Geographic*, March 2, 2015. http://news.nationalgeographic.com/news/2015/03/150302-syria-war-climate-change-drought/ (accessed September 16, 2015).

Wilson, Duff. "The Village That Must Move—But Can't." 2014. Thomsonreuters.com. http://www.reuters.com/investigates/special-report/waters-edge-the-crisis-of-rising-sea-levels/#article-3-grand-designs (accessed August 12, 2015).

World Bank. "Kiribati: Pushing Against the Tide." October 26, 2011. www.worldbank.org/en/news/feature/2011/10/26/kiribati-pushing-against-the-tide (accessed September 6, 2015).

10. Deny, Debate, and Delay

Bast, Diane. "30,000 Scientists Sign Petition on Global Warming." July 1, 2008. Heartland Institute. http://news.heartland.org/newspaper-article/2008/07/01/30000-scientists-sign-petition-global-warming (accessed September 6, 2015).

Benestad, Rasmus E., Dana Nuccitelli, Stephan Lewandowsky, et al. "Learning from Mistakes in Climate Research." *Theoretical and Applied Climatology*, August 20, 2015. http://link.springer.com/article/10.1007/s00704-015-1597-5. doi: 10.1007/s00704-015-1597-5 (accessed August 31, 2015).

Booker, Christopher. "Rise of Sea Levels Is 'The Greatest Lie Ever Told.'" *Telegraph*, March 28, 2009. http://www.telegraph.co.uk/comment/columnists/christopherbooker/5067351/Rise-of-sea-levels-is-the-greatest-lie-ever-told.html (accessed September 6, 2015).

Brulle, Robert J. "Institutionalizing Delay: Foundation Funding and the Creation of U.S. Climate Change Counter-Movement Organizations." *Climatic Change* 122, no. 4 (2014): 681–94.

Clement, Scott. "How Americans See Global Warming—in 8 Charts." *Washington Post*, April 22, 2013. http://www.washingtonpost.com/blogs/the-fix/wp/2013/04/22/how-americans-see-global-warming-in-8-charts/ (accessed September 6, 2015).

Colbert, Steven. "The Word: Sink or Swim." *The Colbert Report*, Comedy Central (video), June 4, 2012. http://www.cc.com/video-clips/w6itwj/the-colbert-report-the-word---sink-or-swim (accessed September 6, 2015).

Congressional Record. Senate. "Science of Climate Change." July 28, 2003. Pages S10012–23. http://www.gpo.gov/fdsys/pkg/CREC-2003-07-28/pdf/CREC-2003-07-28-pt1-PgS10012.pdf#page=1 (accessed September 18, 2015).

Cook, John, Dana Nuccitelli, Sarah A. Green, Mark Richardson, Bärbel Winkler, Rob Painting, Robert Way, Peter Jacobs, and Andrew Skuce. "Quantifying the Consensus on Anthropogenic Global Warming in the Scientific Literature." *Environmental Research Letters* 8, no. 2 (2013): 024024. http://www.skepticalscience.com/docs/Cook_2013_consensus.pdf;doi:10.1088/1748-9326/8/2/024024 (accessed September 6, 2015).

Coolidge, Sheila. "Florida's Citizens Property: Too Big to Fail?" March 19, 2012. Wolters Kluwer Financial Services. Compliance Corner. http://www.wolterskluwerfs.com/article/floridas-citizens-property-too-big-to-fail.aspx?terms=too big to fail (accessed September 6, 2015).

Desilver, Drew. "Most Americans Say Global Warming Is Real, but Opinions Split on Why." June 6, 2013. Pew Research Center.

www.pewresearch.org/fact-tank/2013/06/06/most-americans-say-global-warming-is-real-but-opinions-split-on-why/ (accessed August 13, 2015).

Disaster Relief Appropriations. Public Law 113-2, 127 Stat. 4. 2013. https://www.govtrack.us/congress/bills/113/hr152/text/enr (accessed September 18, 2015).

Gillis, Justin, and John Schwarz. "Deeper Ties to Corporate Cash for Doubtful Climate Researcher." *New York Times*, February 21, 2015. http://www.nytimes.com/2015/02/22/us/ties-to-corporate-cash-for-climate-change-researcher-Wei-Hock-Soon.html?_r=0 (accessed February 25, 2015).

Goldenberg, Suzanne. "Secret Funding Helped Build Vast Network of Climate Denial Think Tanks." *Guardian*, February 14, 2013. http://www.theguardian.com/environment/2013/feb/14/funding-climate-change-denial-thinktanks-network (accessed September 6, 2015).

Guo, Jeff. "The Old Man and the Rising Sea." *Washington Post*, December 2, 2014. http://www.washingtonpost.com/news/storyline/wp/2014/12/02/the-old-man-and-the-rising-sea-2/ (accessed December 26, 2014).

Hasemyer, David. "Documents Reveal Fossil Fuel Fingerprints on Contrarian Climate Research." *Inside Climate News*, February 21, 2015. http://insideclimatenews.org/news/21022015/documents-reveal-fossil-fuel-fingerprints-contrarian-climate-research-willie-soon-harvard-smithsonian-koch-exxon-southern-company?utm_source=Inside+Climate+News&utm_campaign=c6d57d21bf-Weekly_Newsletter_2_22_20152_20_2015&utm_medium=email&utm_term=0_29c928ffb5-c6d57d21bf-327500721 (accessed February 24, 2015).

Humlum, Ole, Jan-Erik Solheim, and Kjell Stordahl. "Identifying Natural Contributions to Late Holocene Climate Change." *Global and Planetary Change* 79 (2011): 145–56.

Intergovernmental Panel on Climate Change. "Climate Change 2013: The Physical Science Basis." 2013. www.ipcc.ch/report/ar5/wg1/ (accessed August 13, 2015).

International Union for Quaternary Research. "INQUA Statement on Climate Change." 2014. http://www.inqua.org/files/iscc.pdf (accessed September 7, 2015).

Jastrow, Robert, to Terry Yosle. February 22, 1991. WAN papers, Accession 2001-01, 60: file label "Marshall Institute Correspondence, 1990–1992." San Diego: University of California, SIO Archives.

Korten, Tristram. "In Florida, Officials Ban Term 'Climate Change.'" Florida Center for Investigative Reporting. *Newsletter*, March 8, 2015. fcir.org/2015/03/08/in-florida-officials-ban-term-climate-change/; *Miami Herald*. March 8, 2015. http://www.miamiherald.com/news/state/florida/article12983720.html (accessed August 11, 2015).

Littlemore, Richard. "What Passes for a Brain Trust at Heartland?" *DeSmogBlog*, February 23, 2012. http://www.desmogblog.com/what-passes-brain-trust-heartland (accessed September 7, 2015).

Lynas, Mark, and George Monbiot. "The Spectator Runs False Sea-Level Claims on Its Cover." *George Monbiot's Blog, Guardian*, December 2, 2011. http://www.theguardian.com/environment/georgemonbiot/2011/dec/02/spectator-sea-level-claims (accessed September 7, 2015).

MacArthur, Ron. "The Other Side of the Sea-Level Rise Debate." *Cape Gazette of Lewes, Delaware*, March 22, 2013. http://www.smalltownnews.com/article.php?pid=4&aid=133592 (accessed September 7, 2015).

MacDonald, Lawrence. "CGD Ranks CO2 Emissions from Power Plants Worldwide." November 14, 2007. Center for Global Development. http://www.eurekalert.org/pub_releases/2007-11/cfgd-crc111207.php (accessed September 4, 2015).

Mann, M. E., C. M. Ammann, R. S. Bradley, et al. "On Past Temperatures and Anomalous Late-20th-Century Warmth." American Geophysical Union. Press release, July 7, 2003. "Leading Climate Scientists Reaffirm That Late 20th Century Warming Was Unusual and Resulted from Human Activity." http://www.geo.umass.edu/faculty/bradley/mann2003a.pdf (accessed September 7, 2015).

Mann, Michael E., Raymond S. Bradley, and Malcolm K. Hughes. "Global-Scale Temperature Patterns and Climate Forcing over the Past Six Centuries." *Nature* 392 (1998): 779–87. http://www.geo.umass.edu/faculty/bradley/mann1998.pdf (accessed October 21, 2015).

McCandless, David, and Helen Lawson Williams. "Climate Change: A Consensus Among Scientists?" December 23, 2009.

InformationIsBeautiful.net 2.0. http://www.informationisbeautiful.net/2009/climate-change-a-consensus-among-scientists/ (accessed September 7, 2015).

Mörner, Nils-Axel. "Rising Credulity: The Truth About Sea Levels? They're Always Fluctuating." *Spectator*, December 3, 2011. http://www.spectator.co.uk/features/7438683/rising-credulity/ (accessed September 7, 2015).

Mörner, Nils-Axel, Michael Tooley, and Göran Possnert. "New Perspectives for the Future of the Maldives." *Global and Planetary Change* 40, nos. 1–2 (2004): 177–82. http://www.sciencedirect.com/science/article/pii/S0921818103001085. doi:10.1016/S0921-8181(03)00108-5 (accessed September 7, 2015).

N.C. Coastal Resources Commission, Science Panel on Coastal Hazards. "North Carolina Sea-Level Rise Assessment Report." March 2010. N.C. Department of Environment and Natural Resources, Division of Coastal Management. http://www.ncleg.net/documentsites/committees/LCGCC/Meeting%20Documents/2009-2010%20Interim/March%2015,%202010/Handouts%20and%20Presentations/2010-0315%20T.Miller%20-%20DCM%20NC%20Sea-Level%20Rise%20Rpt%20-%20CRC%20Science%20Panel.pdf (accessed September 7, 2015).

N.C. Coastal Resources Commission, Science Panel on Coastal Hazards. "North Carolina Sea Level Rise Assessment Report, 2015 Update to the 2010 Report and 2012 Addendum (Draft Report)." March 31, 2015. N.C. Department of Environment and Natural Resources, Division of Coastal Management. http://portal.ncdenr.org/c/document_library/get_file?uuid=dd00328d-67d4-4f39-9e8c-6585cae50577&groupId=38319 (accessed September 7, 2015).

Nuccitelli, Dana. "Global Warming: Why Is IPCC Report So Certain About the Influence of Humans?" *Guardian*, September 27, 2013. http://www.theguardian.com/environment/climate-consensus-97-per-cent/2013/sep/27/global-warming-ipcc-report-humans (accessed September 7, 2015).

Nuccitelli, Dana. "Here's What Happens When You Try to Replicate Climate Contrarian Papers." *Guardian*, August 25, 2015. http://www.theguardian.com/environment/climate-consensus-97-per-cent-2015/aug/25/heres-what-happens-when-you-try-to-replicate-climate-contrarian-papers (accessed August 27, 2015).

Open Mind: Science, Politics, Life, the Universe, and Everything. "Horseshit Power." *Internet Blog*, December 18, 2012.http://tamino.wordpress.com/2012/12/18/horseshit-power/ (accessed August 31, 2015).

Oreskes, Naomi. "The Scientific Consensus on Climate Change." *Science* 306: 1686 and 1 (addendum). December 3, 2004; corrected January 21, 2005. http://cmbc.ucsd.edu/Research/Climate_Change/Oreskes%202004%20Climate%20change.pdf (accessed September 7, 2015).

Oreskes, Naomi, and Erik M. Conway. *Merchants of Doubt: How a Handful of Scientists Obscured the Truth on Issues from Tobacco Smoke to Global Warming*. New York: Bloomsbury Press, 2010.

Payne, Verity. "Rising Incredulity at the Spectator's Use of Dubious Sea Level Claims." December 2, 2011. Carbon Brief. http://www.carbonbrief.org/blog/2011/12/rising-incredulity-at-the-spectator%E2%80%99s-use-of-dubious-sea-level-claims/ (accessed September 7, 2015).

Pew Research Center for the People & the Press. "More Say There Is Solid Evidence of Global Warming." October 15, 2012. www.people-press.org/2012/10/15/more-say-there-is-solid-evidence-of-global-warming/ (accessed August 13, 2015).

Rowland, Christopher. "Researcher Helps Sow Climate-Change Doubt." *Boston Globe*, November 5, 2013. http://www.bostonglobe.com/news/nation/2013/11/05/harvard-smithsonian-global-warming-skeptic-helps-feed-strategy-doubt-gridlock-congress/uHssYO1anoWSiLwov1YcUJ/story.html (accessed September 7, 2015).

Seitz, Frederick. "Global Warming and Ozone Hole Controversies: A Challenge to Scientific Judgment." Washington, D.C.: George C. Marshall Institute, 1994. http://legacy.library.ucsf.edu/tid/pwe37b00/pdf (accessed September 7, 2015).

"Smoking and Health Proposal." 1969. Brown & Williamson; Minnesota Lawsuit. University of California, San Francisco, Library. http://industrydocuments.library.ucsf.edu/tobacco/docs/psdw0147 (accessed September 6, 2015).

Soon, Willie, and Sallie Baliunas. "Proxy Climatic and Environmental Changes of the Past 1000 Years." *Climate Research* 23 (2003): 89–110. http://www.int-res.com/articles/cr2003/23/c023p089.pdf (accessed July 30, 2015).

Soon, Willie, and Nils-Axel Mörner. "Soon and Morner: Sea-Level Rise Data Based on Shoddy Science: Stemming the Tide of Political Fear-Mongering." *Washington Times*, December 17, 2012. http://www.washingtontimes.com/news/2012/dec/17/sea-level-rise-data-based-on-shoddy-science/?page=all (accessed September 7, 2015).

Stenhouse, Neil, Edward Maibach, Sara Cobb, Ray Ban, Andrea Bleistein, Paul Croft, Eugene Bierly, Keith Seitter, Gary Rasmussen, and Anthony Leiserowitz. "Meteorologists' Views About Global Warming: A Survey of American Meteorological Society Professional Members." *Bulletin of the American Meteorological Society*, July 2014. http://journals.ametsoc.org/doi/pdf/10.1175/BAMS-D-13-00091.1 (accessed August 13, 2015).

Union of Concerned Scientists. "Smoke, Mirrors & Hot Air: How ExxonMobil Uses Big Tobacco's Tactics to Manufacture Uncertainty on Climate Science." January 2007. http://www.ucsusa.org/assets/documents/global_warming/exxon_report.pdf (accessed September 7, 2015).

Woodworth, Philip. "Have There Been Large Recent Sea Level Changes in the Maldive Islands?" *Global and Planetary Change* 49, nos. 1–2 (2005): 1–18. http://www.sciencedirect.com/science/article/pii/S0921818105000780#; doi:10.1016/j.gloplacha.2005.04.001 (accessed September 7, 2015).

Zucchino, David. "In North Carolina, a Fight Over Sea Levels and Science." *Los Angeles Times*, June 24, 2012. http://articles.latimes.com/2012/jun/24/nation/la-na-sea-level-20120624 (accessed September 6, 2015).

11. Ghosts of the Past, Promise of the Future

Castedo, Ricardo, Rogelio de la Vega-Panizo, Marta Fernández-Hernández, and Carlos Paredes. "Measurement of Historical Cliff-top Changes and Estimation of Future Trends Using GIS Data Between Bridlington and Hornsea—Holderness Coast (UK)." *Geomorphology*, November 28, 2014. http://dx.doi.org/10.1016/j.geomorph.2014.11.013; doi:10.1016/j.geomorph.2014.11.013 (accessed September 7, 2015).

Coomes, Phil. "Lost Villages." April 11, 2013. BBC News. http://www.bbc.com/news/in-pictures-22025150 (accessed September 7, 2015).

Cracknell, Basil E. *"Outrageous Waves"—Global Warming & Coastal Change in Britain Through Two Thousand Years*. Chichester: Phillimore, 2005.

Duck, Robert. *This Shrinking Land: Climate Change and Britain's Coasts*. Dundee: Dundee University Press, 2011.

Gale, Alison. *Britain's Historic Coast*. Stroud: Tempus, 2000.

Gornitz, Vivien. *Rising Seas: Past, Present, Future*. New York: Columbia University Press, 2013.

Griggs, Gary B. "Lost Neighborhoods of the California Coast." *Journal of Coastal Research On-line*, November 18, 2013. http://dx.doi.org/10.2112/13A-00007.1 (accessed September 4, 2015).

Sheppard, Thomas. *The Lost Towns of the Yorkshire Coast and Other Chapters Bearing Upon the Geography of the District*. London: Brown, 1912. http://www.archive.org/details/losttownsofyorks00sheprich (accessed September 5, 2015).

Further Reading

Websites

Coastal Care www.coastalcare.org

Environmental Protection Agency. "Climate Change." http://www.epa.gov/climatechange/

Federal Emergency Management Agency. "Coastal Barrier Resources System." https://www.fema.gov/coastal-barrier-resources-system; https://www.floodsmart.gov/floodsmart/; https://msc.fema.gov/portal

Federal Emergency Management Agency. "Hazard Mitigation Grant Program." http://www.fema.gov/hazard-mitigation-grant-program

Federal Emergency Management Agency, National Flood Insurance Program. "About the National Flood Insurance Program." https://www.floodsmart.gov/floodsmart/pages/about/nfip_overview.jsp

Federal Emergency Management Agency. "Pre-Disaster Mitigation Grant Program" http://www.fema.gov/pre-disaster-mitigation-grant-program

Rising Seas. "Greenhouse Effect and Sea Level Rise: America Starts to Prepare" http://papers.risingsea.net/

Interactive Maps

Chacko, Sarah. "Bases at Risk." http://global-warning.org/main/installations/

Climate Central. "Energy Infrastructure Threat from Sea Level Rise." http://sealevel.climatecentral.org/maps/energy-infrastructure-threat-from-sea-level-rise

Climate Central. "Surging Seas." http://sealevel.climatecentral.org/

National Oceanic and Atmospheric Administration, Digital Coast. "Sea Level Rise Viewer." http://coast.noaa.gov/digitalcoast/tools/slr/

National Oceanic and Atmospheric Administration. "Sea Level Rise and Coastal Flooding Impacts." http://coast.noaa.gov/slr/

Rising Seas. "More Sea Level Rise Maps: Coastal Elevations." http://maps.risingsea.net/

Index